The **National Computing Centre** develops techniques, provides services, offers aids and supplies information to encourage the more effective use of Information Technology. The Centre co-operates with members and other organisations, including government bodies, to develop the use of computers and communications facilities. It provides advice, training and consultancy; evaluates software methods and tools; promotes standards and codes of practice; and publishes books.

Any interested company, organisation or individual can benefit from the work of the Centre – by exploring its products and services; or in particular by subscribing as a member. Throughout the country, members can participate in working parties, study groups and discussions; and can influence NCC policy.

For more information, contact the Centre at Oxford Road, Manchester M1 7ED (061-228 6333), or at one of the regional offices: London (01-353 4875), Bristol (0272-277 077), Birmingham (021-236 6283), Glasgow (041-204 1101) or Belfast (0232-665 997).

NCC THE NATIONAL CENTRE FOR INFORMATION TECHNOLOGY

Do You Want to Write?

Could you write a book on an aspect of Information Technology? Have you already prepared a typescript? Why not send us your ideas, your 'embryo' text or your completed work? We are a prestigious publishing house with an international reputation. We have the funds and the expertise to support your writing ambitions in the most effective way.

Contact: Geoff Simons, Publications Division, The National Computing Centre Ltd, Oxford Road, Manchester M1 7ED.

Introducing Systems Analysis

Steve Skidmore
Brenda Wroe

PUBLISHED BY NCC PUBLICATIONS

British Library Cataloguing in Publication Data

Skidmore, Steve
  Introducing systems analysis.
  1. System analysis
  I. Title II. Wroe, Brenda
  003   QA402

  ISBN 0-85012-630-4

© THE NATIONAL COMPUTING CENTRE LIMITED, 1988

First published in 1988 by:

NCC Publications, The National Computing Centre Limited, Oxford Road, Manchester M1 7ED, England.

Typeset in 10pt Times Roman by Bookworm Typesetting, Manchester; and printed by Hobbs the Printers of Southampton.

ISBN 0-85012-630-4

Foreword

This book is the first of two texts about the development of computer systems. It presents a series of models and skills that should help the definition and delivery of appropriate, effective, maintainable and flexible information systems. This first book is primarily concerned with *analysis* and what has come to be known as *logical design*. *Physical design* based on these logical models is the main theme of the companion book *Introducing Systems Design*.

The book is primarily aimed at:

— undergraduate and Higher National Diploma and Certificate students undertaking a module in Systems Analysis and Design;

— trainee Analysts studying for professional qualifications or following professional development schemes.

The questions given at the end of most chapters serve to reinforce the point that this is essentially a *study textbook*. A short supplementary teaching guide for lecturers is available from the NCC.

The text concentrates upon the activities of Systems Analysis and Design and the skills and attitudes required to undertake them. It is not concerned with the organisation and management of the tasks. The different management arrangements of Data Processing (DP) or Information Systems (IS) have been explored elsewhere (Keen, 1981). The term 'implementation' is used to denote the delivery of the system, not its program coding. However, the terms 'analysis', 'design' and 'development' are used fairly interchangeably to describe the whole system activity. The different words are employed for textual variation rather than to suggest the job title of the person undertaking the task.

Similarly, the term 'user' is employed in the sense of embracing the clients, operators and victims of the computer system. The term is too standard for its replacement in a textbook.

A case study is used throughout the book to illustrate the models introduced in the earlier parts of each chapter. It must be stressed that the case study is not complete and does not stand on its own. It is used to present a common theme and environment, not as a complete analysis of an organisation. The case study concerns a developing company called InfoSys which offers computer and management training, publishing and consultancy. It has recently been taken over by a large multinational organisation, COMMUNIQUE, which specialises in this area. The Managing Director of the company is Paul Cronin and Jim McQuith is the Seminar Manager. An overview of the company and its operations is given in the Rich Picture of Chapter 2. InfoSys and COMMUNIQUE are completely fictitious companies but their operations illustrate the business systems orientation of the text.

Two aspects must be made clear.

Methodology

The text is non-methodological in the sense that it does not totally adopt any of the proprietary development methodologies currently available in the commercial market-place. There are two main reasons for this:

1 We do not believe that any of the methodologies are applicable in all development circumstances. They tend to be more relevant to large organisations with significant computer resources undertaking complex projects.

2 Consequently, they are not suitable for general Analyst training and education. The approach adopted here is to select techniques from certain methodologies when they suit the purpose at ·hand. The overall aim is to give the practitioner a set of tools which can be selected to suit different circumstances. The principle of this tool-kit approach has been elaborated elsewhere (Benyon, 1987) and will be reviewed at the end of the companion text.

Top–Down Approach

In general, a 'top–down' approach to system development has been preferred. This is a reflection of the belief that systems should be developed in the context of the business requirements and that this can best be achieved by building systems downwards from an understanding of the enterprise's strategic requirements. Information systems must support critical business areas. The practice of computerising 'obvious' applications, such as payroll and accounting ledgers, has often led to a fragmented computer strategy that does not use resources effectively.

The adoption of this approach does create some teaching difficulty. Many students and teachers may wish to leave a detailed examination of the material covered in Chapter 2 to the end of their examination of the whole system activity. Similarly, the perspective covered in Chapter 1 underpins the whole book and should be regularly reviewed as each chapter is read.

Finally, although we have attempted to provide a detailed teaching text, we strongly feel that this should be supported by case study practice. Systems Analysis and Design remains a practical subject with a significant tangible end product. Teaching Analysis without context can be both unrealistic and unfocused.

References

Benyon D, Skidmore S, Towards a Tool Kit for the Systems Analyst, *The Computer Journal,* vol 30, no 1, 1987

Keen, J, *Managing Systems Development,* John Wiley, 1981

Acknowledgements

We would like to thank colleagues at Leicester Polytechnic for commenting on earlier drafts of this book. We are particularly indebted to David Howe, Michael Martinek and Gillian Mills.

Contents

1   Systems Development: A Perspective

1.1   INTRODUCTION

This chapter provides a brief guide to the tasks of systems development. It begins by placing these in the context of the Systems Life Cycle – a common model for illustrating the progression of analysis and design tasks (see Figure 1.1). Criticisms of the scope of this model are then noted and consequently a number of extensions proposed. The tasks of this larger model are then introduced by reference to the relevant chapter of this book. This is summarised in Figure 1.2, which effectively acts as a 'road-map' to the text, and can be used for selective reading and revision.

The linear progression of tasks described in the extended model is then contrasted with a more iterative approach to development called Prototyping. The philosophy and implementation of Prototypes is discussed and its place in the development model is briefly described.

1.2   THE CONVENTIONAL SYSTEMS LIFE CYCLE

The Systems Life Cycle model (Lee, 1978) is a convenient place to start an examination of what an Analyst actually does. This model (see Figure 1.1 for our version) suggests a number of stages:

The Feasibility Study. An examination of the current operational systems and a brief consideration of alternative ways of computerising these tasks. This comparison is undertaken in the context of economic, technical and operational issues, culminating in a Feasibility Report which recommends a possible solution and comments on whether detailed analysis should commence.

Detailed Systems Analysis. Once the scope of the study has been agreed, a detailed investigation is undertaken of the operations of the current system and the requirements of its successor.

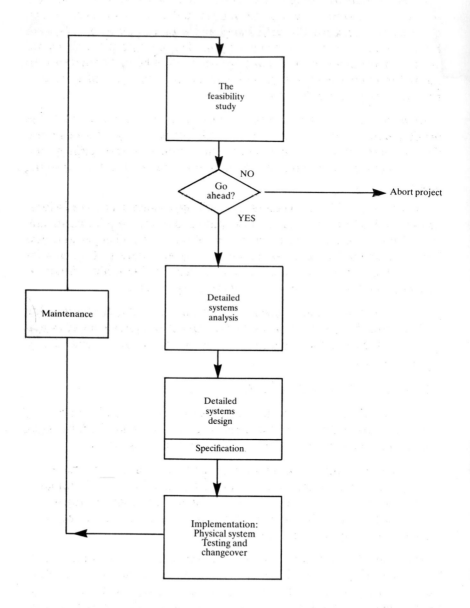

Figure 1.1   The Systems Life Cycle

Detailed Systems Design. The design of a paper specification of a replacement system detailing the format and content of files, inputs, outputs, etc. This usually culminates in the presentation of a System Specification to the user showing the intended contents of the delivered system. Acceptance of this leads to a specification being handed over to the programming team for the construction of a suite of software that produces the promised results.

Implementation. The testing of programs and systems and the development of supporting manuals and documentation. Implementation also includes the phasing in of the new system using an appropriate management method and the organisation and conduct of training courses.

Maintenance. The implementation of amendments and omissions. These are likely to increase as time passes and new requirements and possibilities emerge as the environment of the system changes and new hardware and software opportunities present themselves. This culminates in the consideration of a new system, a Feasibility Study is commissioned, and the development cycle begins again.

This textbook includes, but does not embrace, the traditional life cycle approach. The practice and environment of systems analysis and design has changed dramatically in the past few years and this is reflected in this text.

1.3   EXTENDING THE MODEL

One of the criticisms of the life cycle approach is that it fails to properly address the context in which business systems are developed. Discussing the failure of classic tools of systems analysis, Er states that:

> The Systems Analyst seldom questions why such an information system should be designed in the first place, and what management activities it tries to serve. As a consequence, an efficient information system may be implemented for serving a less effective management system. (Er, 1986.)

The business justification of computer systems is too important to be left to chance. Hence this book begins with an examination of the role of information systems in the context of the enterprise. Every organisation faces at least two fundamental problems when considering computerisation:

— identifying relevant applications;

— giving appropriate priority to these relevant applications.

This can only be done successfully if the business contribution of competing applications is properly understood. Chapter 2 provides two frameworks for gaining such an understanding.

Candidate applications for computerisation may now be subjected to a detailed Feasibility Study, and this is examined in Chapter 3. In most respects this follows the lines of the conventional life cycle model, except that the business orientation of the selection simplifies the cost/benefit analysis as well as removing the burden of justifying *why* that particular area should be computerised.

Chapter 4 looks at some of the social skills needed by developers to successfully analyse, design and implement systems. There is a trend to present systems development as a technical skill; the construction of a set of increasingly complex models that will guarantee success. This is probably a reaction to past conventional training which tended to overstress the personal characteristics of the Analyst, placing too much emphasis on interviewing, conversational and presentation skills. There needs to be a balance between technical modelling and interpersonal skills with the latter seen as supporting the former.

Chapter 5 presents a reasonably conventional approach to the analysis of present operational systems. Some proprietary methodologies, in rejecting the life cycle approach, have either reduced such analysis or introduced different modelling tools. But there is an element of 'throwing the baby out with the bathwater' in doing this. We feel that an understanding of the current information systems that support the organisation gives a fundamental insight into the design requirements of any successor. In addition, it also represented one of the strengths of the life cycle approach and some of its diagramming techniques are demonstrably better than certain of their suggested replacements.

However, there is a problem of moving from an understanding of the present system to the design of a replacement. Relying on current procedures can undoubtedly lead to designs that fail to harness the power of the computer and lead to underperforming 'computerised manual systems' whose restricted scope reduces their chance of success.

This transition from analysis to design has been recognised by the growth of *logical modelling* – techniques that show the logical information system requirements stripped of their administrative arrangements and physical trappings. These models have had a significant impact on the way that computer systems development has been taught and practised. The following three chapters examine complementary logical models.

Chapter 6 introduces Data Flow Diagrams – a process-driven technique that highlights how data is transformed, stored and used as it progresses through the information system. The data flow diagram is a central technique of most of the Structured Methodologies, although notation and nomenclature vary from vendor to vendor.

Chapter 7 examines a technique that takes a more static view of the data. Data Analysis has played an important role in the design of databases, a subject poorly dealt with by a life cycle model geared to the design of conventional files. Furthermore, the data model produced by data analysis provides an important insight into the data structures required and maintained by the organisation. This data-driven view complements the process-driven perspective of the data flow diagrams.

Chapter 8 introduces two important supporting tools that will ease the transition into systems design. The first, the Data Dictionary, provides a facility for the management of the data resource. It is emerging as not only an important *documentor* of systems, but also as a very active tool in the whole system development spectrum. Secondly, the Entity Life History is presented as a technique that can bind together the data and process-driven models. Its emphasis on time and sequence provides a third important logical perspective required for the design of a replacement system.

Finally, Chapter 9 briefly summarises the logical basis of the system design activity discussed in our companion text.

Figure 1.2 describes the succession of models used in this text.

1.4   OBJECTIONS TO THIS MODEL

Another major criticism of the life cycle model is the time-scale associated with its linear progression of activities. The gap between specification and delivery is often so long that requirements change dramatically in this time. This leads to the delivery of systems that 'were required two years ago'. Also, the Specification itself is often not

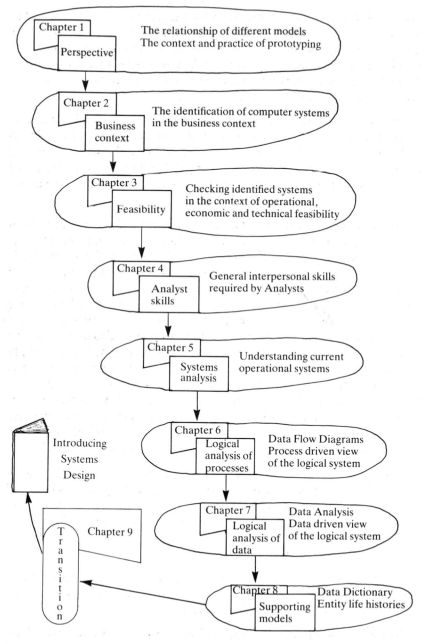

Figure 1.2   A 'Road-map' of this Text

understood by the users responsible for agreeing it. It is often couched in computer terms described at such length that there is little chance of the user really being able to assess if it actually represents his requirements. Consequently, many delivered systems have to undergo changes that reflect misunderstandings and altered circumstances. This is often euphemistically called 'maintenance'.

An alternative approach, known as Prototyping, has claimed advantages in most aspects of the life cycle. Prototypes are a first attempt at a design which is then extended and enhanced through a series of iterations. Prototyping stresses the early delivery of an incomplete, but working, system and the use of Prototypes may be valuable at various stages of the life cycle.

The nature of a Prototype may be illustrated with an example from requirements definition, a task undertaken in the detailed systems investigation. A perceived problem associated with the specification approach is that it presumes the users and operators are able to state their requirements in advance. This seems rather optimistic. In many instances it is unclear to a user how a system may help him, either because the role of the computer is not understood, or because the information needs are unclear. Most users cope with this uncertainty by 'asking for everything' so that amongst the data the system eventually produces are the nuggets of information that they actually require. Thus it is difficult for most users to clearly envisage what they want and how they can use it until they are able to experiment with a tangible system. So a simple Prototype designed to accommodate broad needs, together with possibilities suggested by the designer using experience gained in other projects, may be used to define requirements more accurately.

Prototyping is an increasingly popular method of developing systems and it has clear precedents in other engineering disciplines and activities. It is worth examining in more detail before attempting to place it in the perspective of the models introduced in this book.

1.4.1 Prototype Philosophy

The Prototype is a live, working system and not just a paper-based design. Users can test its operations and explore its facilities and so do not have to rely upon written descriptions.

Prototyping is an iterative process. The first system is built around the user's basic requirements and refined in the light of comments and difficulties (see Figure 1.3). Thus the system passes through a number of iterations until it becomes an acceptable reflection of the user's requirements. At this stage the designer has three options:

— refine the Prototype into a final running system: this may require some development of error trapping and recovery routines;

— recode certain sections of the Prototype to make them more efficient;

— recode the entire system: the Prototype may have been developed with the prime aim of refining user requirements. The software used to do this may turn out to be completely inadequate beyond a certain volume of transactions. Thus recoding is required to permit the phasing out of the Prototype.

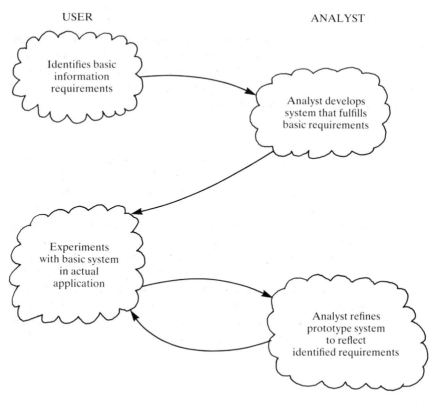

Figure 1.3   The Prototyping Approach

It must be clear that, for such a development method to work, the system has to be written in very powerful and flexible software that permits programs to be created very quickly as well as allowing easy coding of extensions and amendments. Conventional high level programming languages (such as Cobol and Basic) are not suited for this type of development. Prototyping requires powerful software tools that allow the relatively inexpensive building of systems which might eventually be discarded.

1.4.2   Prototype Tools

The tools chosen for Prototype development must, above all, permit the quick development of working systems. Four candidates may be identified.

Application Packages

It may be possible to develop a demonstration system using an appropriate package and to let the user identify problems, possibilities and opportunities using the package as a yardstick. It is often easier to say what is inadequate about, say, a production control package, than it is to define requirements in the abstract.

Program Generators

Program Generators have been available for a number of years and a variety of different types may be identified. One sort uses a question-and-answer English dialogue to produce the program logic which it then encodes in a high level language such as Basic or Cobol. A different type adopts a screen-based approach where the designer effectively 'paints' the screen he wishes to be displayed by typing directly on to it. Once satisfied with the display, he may invoke the program generator which automatically produces the code required to produce that screen. It may also produce validation routines stopping only to request what type and range of data is required in a certain field and what error message should appear when the user makes a mistake. These may be produced directly if the Program Generator has access to a data dictionary.

Reusable Code

Many systems are conceptually similar. Tasks frequently reappear –

menu design, password protection, print routines, date checking, etc. It is possible to build up a library of well-proven, well-documented routines which may be plugged together to make up a system. The content, not the logic, will need changing and some patching will be required to make a complete system. However, access to a store of well-tested standard modules should ease development and maintenance.

Fourth Generation Languages

It is the emergence of Fourth Generation Languages (4GLs) that has largely permitted the adoption of the Prototyping approach. Languages may be seen as passing through three previous generations:

First Generation. Machine Code. Instructing the machine through direct binary code. Closely associated with the architecture of the host processor. Complex to write, read and debug.

Second Generation. Assembler Languages. Uses symbolic codes. Machine instructions given by mnemonic alphabetic codes. Easier to understand although still closely allied to machine architecture.

Third Generation. High level languages such as Fortran, Cobol, PL/1 and Basic. Written in procedural code. Largely independent of the hardware architecture permitting portability. Much easier to use.

There is no agreed definition as to what constitutes a Fourth Generation Language. It might be safer to suggest that it consists of everything which is not in the first three categories! In general such products are marked by:

— non-procedural programming code;

— a simple query language;

— centred around a database.

Such software should provide a whole application environment for the system developer. Martin (1982) provides a checklist of desirable features for 4GL evaluation, and this is summarised below:

— Centred around a relational database.

— Links to other proprietary databases and other non-database files. This will permit gradual transition to the new development strategy.

— Integrated and active data dictionary.

— Simple Query Language. May use a syntactical structure and/or a Query by Forms.

— Integrated Screen Design Tool.

— Dialogue Design Tool. Including generation and manipulation of business graphics.

— Report Generator.

— Procedural Coding facility. This may be done directly through a conventional language (say Cobol) or indirectly via a code Design Aid (such as an Action Diagrammer).

— Non-Procedural Programming Code.

— Spreadsheets and graphics.

If the 4GL is to provide a complete development facility for the professional programmer, Analyst and end-user it must clearly have a range of tools to accommodate disparate requirements and skill levels. It must also dovetail with the past development strategy of the organisation and the systems developed under that strategy. 4GLs that require massive rewrites of current operational systems are likely to extend the development backlog, not reduce it.

The range of software that is either marketed or perceived as a Fourth Generation Language is demonstrated by a survey (Lobel, 1984) which lists about 66 products, ranging from the relatively sophisticated (such as Oracle and Ramis) to the rather simplistic (for example, FMS-80 and Autocode).

Not all research on Prototyping has used 4GLs. For example, one of the most quoted research papers used Pascal for both Specification and Prototyping teams (Boehm, 1984). Similarly, 4GLs can (and are) being used in a life cycle environment, particularly to cut down programming time. But there is clearly a case for linking the two, in that 4GLs facilitate Prototyping.

1.4.3 Prototyping: Summary

Prototyping may be viewed as a different response to the problems perceived in the conventional Systems Life Cycle. It stresses the delivery of an early working system that is gradually refined through user experience and comments. It is likely that Fourth Generation development tools will be used to construct this system as these permit the required speed and flexibility.

But it must be stressed that Prototyping is not a panacea. Alavi (Alavi, 1984) listed four perceived advantages and four disadvantages. Amongst the latter were two that are directly addressed by the more linear approach to development:

Management and control. The absence of definite phases makes planning and budgeting very difficult. Documentation and testing also tends to be overlooked.

Size. It is very difficult to Prototype large information systems. Prototyping may lead to inappropriate applications and incompatibility.

1.4.4   Prototyping in our Development Model

This chapter began by suggesting deficiencies in the model of the conventional Systems Life Cycle. These problems were addressed by extending the task to include logical modelling. But this has the drawback of lengthening the whole development cycle. One of the criticisms of some of the methodologies that have adopted this approach is that they are time consuming and bureaucratic. They have so many models and tools that they have become complex and cumbersome.

Prototyping tries to deal with complexity in a different way. It chooses to define a solution through a series of iterations which move progressively closer to the user requirements. But in doing so, it may lose perspective and performance and hence become counter-productive.

Edward Yourdon, amongst others, believes that 4GLs and Prototyping packages have been oversold, and that sound foundations have been sacrificed for short-term expediency (Yourdon, 1986). The emergence of Computer Aided System Design tools (often referred to under the generic term Analyst Workbenches) promises to reduce the modelling time of large systems. This will remove one of the large stumbling blocks of logical modelling and perhaps swing the pendulum away from Prototyping.

For the purpose of this book we have chosen to introduce the *philosophy* of Prototyping in this chapter and then to give examples of its use at different stages of the overall task. For instance, the potential use of a Prototype in defining user requirements is presented in the appropriate place – Chapter 4. Similarly, the design and scope of Prototypes will be considered in the companion text. Our intention has been to thread Prototyping throughout the book, rather than to treat it as a completely separate issue. The same approach has been taken with the automation of the various activities.

However, the sequence of this book is one that is imposed by the medium of text and the need for a clear structure in learning. It is important that this is not regarded as a blue-print for all systems development. In certain circumstances a subset of these techniques or use in a different order may be more appropriate. This is central to the Tool-Kit approach and an issue that is revisited at the end of the companion text *Introducing Systems Design*.

1.5 SUMMARY

This chapter has:

Introduced the tasks of the Systems Life cycle model.

Extended the model to embrace project identification and selection.

Introduced the concept and techniques of logical modelling and added these to the linear model.

Presented a diagrammatical summary of this extended model and indicated the appropriate chapter for each task.

Explained and considered an iterative approach to development called prototyping and explained how this fits into the overall approach of this text.

2 Information Systems: Strategic Issues

2.1 INTRODUCTION

This chapter looks at the early stages of a project, examining some of the issues that will be encountered in project initiation and selection.

2.2 PROBLEM IDENTIFICATION AND SELECTION

Every organisation faces two fundamental problems when it considers the role of computers:

— It has to identify opportunities for computerisation.

— It needs to give these opportunities appropriate priority.

The Analyst has an important role to play in these two activities and this will require a certain amount of professional detachment. It is important not to jump to conclusions when establishing areas of the enterprise that might benefit from computerisation. Payroll, for example, is a common and relatively simple application, but there is no reason why this should necessarily be computerised in every organisation. The fact that something can be computerised does not mean that it should be. Computer systems that aid the organisation's aims and prosperity should receive priority. A simple, slightly fictionalised, example illustrates this.

LDF Engineering was a small engineering firm based in the Midlands. It appointed a keen and aggressive accountant who set about the task of acquiring the firm's first computer to support his planned computerisation of the accounting ledgers and payroll. The hardware and software selection was performed meticulously and an appropriate system was purchased and installed. Just over one year later the firm went into voluntary liquidation. At the post mortem it was felt that overstocking of certain product lines and poor production planning and control had been major contributors to the company's decline. Yet these are applications suited to computerisation! In retrospect, the firm appeared to have successfully computerised the *wrong* system. If better stock and production controls had been applied through using a computer system they may have resulted in something that the accounts application could not offer – the organisation's survival.

The selection of relevant applications is undoubtedly much easier in organisations where information systems planning is seen as an integral part of the enterprise's overall planning effort. In such instances the computer systems are recognised as a valuable resource within the organisation's total activity, not as an appendix to it. If enterprises fail to have any overall business planning, then it is difficult to formulate a coherent information system strategy. This corporate view of computing is central to the Effective Computer approach developed by Grindley and Humble (Grindley, 1973) and based upon *Management by Objectives*. Their book was written in response to observations of many projects where the computer failed to satisfy the expectations of managers, a problem that still exists over a decade later. They observe that:

> The only valid objective for computers is to assist in achieving defined business improvements which would be impossible or uneconomic without the computer.

Thus the computer is viewed as a resource to help the *doers* and *makers* of an organisation achieve their profit or service objectives. Benefits are seen to be of two fundamental types – getting more for the things that are provided or sold (selling more, selling at a higher price) or reducing the costs of making and selling them (carrying less stock, making production more efficient). However, as Grindley and Humble note, only in a small number of firms (those selling computer software, services or facilities) can the computer department directly generate income.

In general, they claim that:

> Those who run the computer do not usually have the responsibility for selling, pricing, or production efficiency. It is not open to them to obtain benefits directly. Their role, therefore, must be to help those who can achieve benefits.

Thus, information systems must support the enterprise's doers and makers, those who are given defined business objectives which they must achieve.

2.3 STRATEGIC PLANNING: AN INFORMATION SYSTEMS PERSPECTIVE

If appropriate system development is easier in an organisation with a coherent plan, then it is important that every Analyst should have some insight into the Strategic Planning activity. In this way he can understand its implications for information systems (IS) development as well as (at senior levels) contributing to the initial formulation of the plan. Every manager is part planner, and omitting Information System managers from the overall planning task is likely to lead to inappropriate IS activity and missed opportunities.

"Planning is the design of a desired future and of effective ways of bringing it about." (Ackoff, 1970.)

A corporate appraisal of the enterprise's current strengths and weaknesses is an important first step in the planning process. It sets a baseline for planning as well as providing an opportunity to critically examine established beliefs, operations and practices. It may also directly identify means for profit improvements.

Hussey (Hussey, 1982) identifies 10 major aspects of a corporate appraisal and some of these have important IS implications:

1 *Trends of results.* The company's historical pattern of performance. This will cover trends in profit, sales, capital employed, as well as all the various financial ratios that can be applied to measure efficiency and performance. This will give an overall impression of whether the company is improving or worsening its position. Declining or troubled companies may impose tighter constraints on IS development and expect quicker returns. Alternatively, and perhaps more dangerously, they may not invest enough in system development (preferring to attack what they consider to be the 'real' – ie immediate – problems), and so miss technological opportunities.

2 *Sources of profits.* This will consist of a detailed examination of profitability and prospects. Such a study aims to identify the actual 'breadwinners' of an organisation and this may lead to product variety reduction, elimination of certain products, etc. If this study is not undertaken then IS development time and resources may be wasted upon products and problems which are not important in the production of profit or the delivery of services.

mustn't put profits at risk during upgrade.

3 *Risk.* An examination of the risk associated with the company's sources of profits. This is of particular significance to enterprises that are very dependent upon one product or a few customers or suppliers for trade. Effective information systems are critical in high-risk markets or circumstances, so that the very damaging effect of detrimental changes can be quickly identified and corrective action attempted.

The importance of marketing

4 *Manufacturing activity.* This considers the product processes and identifies possibilities of new production opportunities. The traditional commercial basis of much IS development has undoubtedly led to a lack of consideration of production and design opportunities.

5 *Rationalisation of resources.* A study of whether plants, depots and buildings cannot be amalgamated in some way. This can be applied to computer facilities. There may be much to gain by using such resources in a more cost-effective manner.

benefits to company combine depts

6 *Organisation and management.* The organisational structure of the company and its strengths and weaknesses. This will include problems that might appear in the future such as retirements and difficulties of succession. Managerial capabilities have to be assessed and their strategic implication understood. As Hussey comments:

> There is very little point in deciding to launch a range of new products if it is known that the marketing manager is incapable of making a success of the venture. (Hussey, 1982.)

The importance of corporate structures to the IS developer is considered in the next two chapters. However, Hussey's point is valid at the strategic level of project selection. An inventory management system does not, in itself, lead to a reduction in stock levels, it merely facilitates it. An incompetent stores manager might only make poor decisions faster and at greater cost if supported by a newly installed information system.

The general 'climate' of the company is also important. Computer systems usually bring significant changes and the Analyst must assess how those changes are likely to be viewed by employees and what mechanisms for implementing change already exist in the company.

7   *Financial resources.* An assessment of the company's capital and liquid resources and current and projected cash flow. A company with a very high debt looks at the future very differently than one with a 'cash mountain'. Economic feasibility (discussed in the next chapter) is critically affected by this issue. The financial position determines how systems are funded, when they will be paid for and, most importantly, how much is available for their development. A clear understanding of the financial position is required if projects are to be sensibly assessed and selected.

8   *Corporate capability.* A review of the strong and weak points of the corporate operation: 'What we are good and bad at doing and why'. This will again aid project selection.

9   *Systems.* An appraisal of the strengths and weaknesses of the operational systems that support the company. This will not only look at current efficiency, but also the effect of likely changes:

— Can the order processing department cope with the projected increase in orders for this product?

— What are the picking routines in the warehouse?

— How are marketing promotions decided?

This assessment is clearly of vital strategic importance to the IS section. It may encompass systems already under its jurisdiction and subject their performance to searching corporate review. Also, it is likely to identify possibilities for further computerisation and incorporate these in the business objectives of the resultant Strategic Plan.

10 *Use of resources.* Identifying how the various resources – people, money, plant, etc, are brought together to produce the company's products. This will include a consideration of whether the adopted pattern is the most effective one. Are the right people employed in the right place? Is the product emphasis correct? and so on. One of these resources is information systems and such an appraisal may lead to radical changes in organisation – for example, from centralised to decentralised processing.

Thus the corporate assessment is a wide-ranging examination of the company's position. Information systems development will be affected by all aspects of this appraisal and fundamental insights and 'ground rules' will be established. It provides a statement of the current position of the enterprise.

The next step is to plan the movement of the company from its present position to its desired state. The desired state will be described in terms of objectives.

2.3.1 The Importance of Objectives

Hussey suggests a four-part hierarchy of objectives:

1   *A primary, or profit, objective of the business set before consideration of strategy.* This is an overall target for the company established in advance of how that target is to be achieved. Examples might include:

    — To treble after-tax profits from £250,000 to £750,000 by the end of the fifth year of the plan.

    — To provide a £50,000 income package for each of the partners by the April of the second year of the plan.

    Hussey argues that most primary objectives will be established in terms of profit because, even if the company has other objectives, it must ultimately make a profit or else it will sooner or later cease to exist.

2   *Secondary, mainly narrative, objectives again set in advance of strategy.* These are descriptions of the intended corporate identity. It is a picture of the company's aims in terms of the nature and scope of the business, its trading and staff relations and intended developments, expansions and locations. In this respect it is a statement of how the Chief Executive, Chairman or Directors perceive the future of the company. The narrative objectives are a method of trying to communicate and share that perception with the rest of the company.

3   *Goals which are targets derived from strategy with time-scales.* Goals are milestones on the way to the overall target set prior to strategy formulation. A network of compatible goals lead to the achievement of the primary objective. These goals are derived from strategy and hence can only be properly defined after that strategy has been determined. Goals might be established for:

— the capture of a certain market share;

— an absolute value of sales;

— a minimum figure for customer complaints;

— a cost reduction target;

— a minimum time for answering an emergency call; *Goals for Comms Lines*

— a date by which a product launch must take place.

4 Finally, *Standards of Performance (often identical with goals) are given to particular individuals.* This will often be achieved by splitting up the corporate goals. For example, an overall sales figure may be divided into territories; a profit target, into departments, etc. A set of Performance Standards will be compatible with a goal which, in turn, will complement the other goals contributing to the overall primary objective of the company. This gives a direct link between the expected performance of the individual and the corporate task.

Grindley and Humble suggest an approach to computerisation centred around goals and objectives.

The second stage of the Effective Computer approach establishes corporate objectives and identifies how computer systems might contribute to achieving them. A series of examples are given, two of which are repeated below.

Corporation: Aircraft spares *R/R*

Key area: Market share

Objective: To increase sales volume by 40% *(business)*

Contribution: To install an on-line terminal in customer's premises showing our stock and delivery position and price.

Corporation: Carton manufacturer

Key area: Productivity

Objective: To reduce waste on box manufacture by 8%

Contribution: To implement an order analysis system for determining optimum width and run size for corrugated paper.

The strength of this business approach to information system development is that it places the computer system into the perspective of corporate objectives; this is appropriate because these are the things that actually matter. The computer is seen as an aid to achieving some new business goal, one that would be difficult, costly or impossible to achieve without computerisation.

Note that this approach differs from those which simply attempt to perform some existing system better (cheaper, faster, more accurately, etc). (Grindley, 1973.)

Thus information system strategies have been identified in a 'top-down' way, moving from an overall corporate target to individual goals which might be achieved with the aid of a computer system. This should ensure a unity of purpose impossible with piecemeal and unfocused system development. Three important points about objectives must be made before this approach is applied to the case study:

1   *It is important that the objectives are quantified.* This forces realistic target setting and agreed measures of success. An objective to 'reduce inventory holding' is successfully achieved if only one less item is carried! However, an objective to reduce inventory by 5% by January 1987 makes the criteria of success much more specific. Time constraints are also important for objectives, so that they not only indicate what should have resulted but by when it should have taken place.

    A main difference between companies which practise effective formal planning and those which have a more traditional approach to management is that the planning companies are not satisfied with words alone. Much more meaningful is a specific quantitative statement of what profit is required. (Hussey, 1982.)

2   *Objectives can be set for non-profit making organisations.* These might include:

    — all emergency calls will be answered in 2 hours;

    — no patient will have to wait more than 30 days for throat surgery;

    — at least 80% of our students will pass the Level 2 examinations.

3   *Objectives are set for a particular business goal, not for the computer's*

contribution. Any attempt to partition objectives, or indeed success, between the computer and the non-computer part of the corporate effort is doomed to failure. The important matter is to achieve objectives, not to squabble about who, or what, achieved them.

2.4   STRATEGIC PLANNING AT INFOSYS

Primary Objective set Prior to Strategy

This is set by the new parent company – COMMUNIQUE. They wish all their companies to increase pretax profits by 25% within three years.

Secondary, Narrative Objectives

There will clearly be a fairly long list of these secondary objectives. Two illustrative examples arc:

— The company is operating in an increasingly competitive market-place where 'smallness' in company size is not a favourable marketing attribute. Hence it is the intention of the company to stress the multinational nature of the organisation and to promote the European COMMUNIQUE image. Customers and employees will be made increasingly aware of the fact that they are working for, or buying from, a COMMUNIQUE company rather than InfoSys.

— The company wishes to widen its service base. This will be achieved by identifying and developing products and services that will not only complement services currently offered, but will also lead to 'Added Value' sales opportunities for the established range. The current reliance of the company on a relatively narrow service base, with an increased number of competitors, has led to the recognition of the need to diversify.

Goals

The primary objective established by COMMUNIQUE represents the overall target of the Strategic Plan. The management of InfoSys is now charged with the responsibility of drawing up a series of goals that will contribute to the achievement of this primary objective. Their plan will clearly be a wide-ranging document covering most aspects of the company and some of these issues will not be of immediate relevance to information systems development. Furthermore, it is only necessary to give a flavour of the planning document for the current purpose of illustration. Hence, only two goals are looked at in more detail.

Goal 1: To increase seminar income by £300,000 per annum within three years.

Discussion

This increased target has been agreed by Managing Director Paul Cronin with the Seminar Manager Jim McQuith. The Corporate Appraisal has identified that seminars have traditionally been an important source of profits, but that the section has not been allocated the resources needed to meet the perceived demand. The Strategic Plan has suggested a major expansion in the seminars section, employing three Seminar Executives and two further administrative assistants. Individual standards of performance have been specified for these posts (see Figure 2.1).

Functional area	Responsibility	Year 1	Year 2	Year 3	Year 4	Year 5
Strategic Planning Training	Jim McQuith	70	75	75	75	75
Systems Development Training	*Seminar Executive 1	10	50	75	75	75
Microcomputer Training	*Seminar Executive 2	10	50	75	75	75
Software Engineering Training	*Seminar Executive 3	10	50	75	75	75
	Total	100	225	300	300	300

*Not yet in post.

Figure 2.1   Goals Analysis. Goal 1: Target Growth (Figures in £000s)

However, it was also acknowledged that these new staff could only be effective if they were supported by a more efficient information system. The section was currently profitable despite the rather chaotic nature of its clerical methods. A number of specific issues were highlighted in the Strategic Plan.

— The need for accurate information on seminar bookings. The associated tasks of material production, accommodation booking and catering were hampered by the absence of accurate, timely information.

— Greater quality control in seminar delivery. The seminars were not evaluated properly. There was no specific information on seminar popularity, lecturer performance, booking trends, customer characteristics, etc.

— The marketing of the seminar programme was hampered by the absence of any systematic record keeping. Information about delegates was held in separate course folders and sometimes manually extracted for special mailshots. Enquiries were seldom recorded in any usable fashion. The success of the advertising of the seminar programme in a variety of publications and magazines was not evaluated in any way.

Jim McQuith agreed that the new expanded seminars section would have to be supported by a better information system. A Feasibility Study was commissioned to look into this in more detail.

Thus an important information system contribution has been identified stemming from a required business goal. This is a goal that cannot be achieved by using the current information system.

Goal 2: The second goal deserves a longer introduction. One of the areas covered in the Corporate Appraisal was *Risk*. The point has already been made that InfoSys recognises its weaknesses and intends to adopt a number of policies to counter them. Two have been explicitly mentioned:

— To give InfoSys the image of a larger company by changing its corporate presentation to COMMUNIQUE.

— To diversify the product and service range.

A way of achieving the latter was identified during the Corporate Appraisal. A planning group looking at the performance of the publications section requested an analysis of sales of a sample of 20 books produced by the company two years ago. The books were selected so that they represented different technical areas. They had also, without exception, come to the end of their publication life – further sales were expected to be insignificant. The results of their survey are given in Figure 2.2. A number of discussions were held about how sales might be increased. There was general dissatisfaction with conventional methods of distribution. Many booksellers did not recognise that they were dealing with short-term products and so failed to promote them sufficiently. Catalogue compilation and distribution

Copies sold	<2000	2000–3000	3000–5000	5000–8000	8000–10,000	10,000+	Total
Frequency	4	6	4	3	2	1	20

Figure 2.2   Book Sales Analysis

was also criticised. The catalogues were generally thought to be unattractive, too inflexible and the response of catalogue recipients was not analysed in any way.

After much discussion it was agreed that "a method of distribution should be adopted that permitted fast, cost-effective distribution to a well-defined market group". The agreed solution was summarised in a further goal:

> That a Mail-Order Computer Book Club should be established to offer COMMUNIQUE and other titles at discount prices. This Club will be established in Year Two of the Plan and will turnover £100,000 per annum by the end of Year Four of the Plan.

One of the important implications of this goal is that the company will need to set up an efficient mailing system to support its direct selling. This will be necessary for, amongst other things, providing goods to customers, verifying that customers are fulfilling their terms of membership, and analysing mailshot responses and product sales.

There are actually many information system opportunities associated with this planned service diversification. But it is useful to focus briefly upon the mailshot management and analysis as this is perceived as critical to the new venture and also to the efficiency of other parts of the organisation (the seminars section has already been cited). Providing an effective service whilst maintaining low overheads demands a supporting information system that will:

— accurately identify targets for mailshots;

— produce mailing labels in a number of formats;

— analyse the success of each mailshot;

— process mail orders and maintain inventory records for the ordered goods.

The details of the mailshot system can again be discovered in the later phases of analysis (see Chapter 6). What is important is the *process* of *identifying* this application. The Corporate Appraisal has highlighted a *risk*. Management has responded to that risk by planning product diversification within the constraints of current markets and expertise. An objective has emerged from that response which requires a significant amount of information system support.

Thus the business orientation of project selection has again been demonstrated. Two possible application areas have emerged that contribute towards goals which, in turn, serve the overall corporate objective. In some respects both systems are slightly unconventional and may not have been automatic first choices if project selection did not have the recommended corporate focus. More standard applications – such as payroll – are easier to recognise but they may not be worthwhile, or demand the priority that they often receive. Only a business orientated approach can determine such a choice.

2.5   STRATEGIC PLANNING: SUMMARY

Viewing information system development from the strategic business perspective is both useful and valid, but it has a number of possible drawbacks:

— It may provide a rather clinical and idealistic image of the current and future states of the organisation. Enterprises tend to consist of many interrelationships, fired by complex webs of operations, conflict and deceit. The actual enterprise is much messier than the rather sanitised version that often emerges from the planning document.

— It represents a managerial perspective of the organisation. This is clearly an important one, but it is not the only valid viewpoint. Not all employees concerned with information system change will share the management's perspective of the company and the Analyst must be aware of this. Furthermore, management's view of reality may differ considerably from what is actually taking place. Descriptions of the same system given by managerial and operational staff may differ alarmingly in both detail and perspective.

— A plan does not greatly help the Analyst come to grips with the current state of the company. Organisations are often very complex and some method is usually needed of understanding and simplifying this reality.

[handwritten margin note: my project must force uses to 'provide strategic plan]

— Many companies do not have a Strategic Plan, and of those that
  do, many do not consider information system development in its
  organisational perspective. In such instances there is no obvious
  starting place for project selection and priority.

Consequently we need to employ a framework that might concede the
preceding points as well as giving an insight into strategic planning
initiatives. The Soft Systems methodology of Checkland (Checkland,
1981) is a candidate framework.

2.6   SOFT SYSTEMS METHODOLOGY

The Soft Systems framework is concerned primarily with tackling the
ill-structured problems of the real world and suggesting solutions that
may, or may not, include computers. It attempts to tackle the fuzzy and
complicated problems of organisations where there often appear to be
insurmountable obstacles in defining the problem, let alone solving it.
The framework was developed and presented by Peter Checkland
(Checkland, 1981) and variations appear in Wilson (Wilson, 1984) and
Wood-Harper, Antill and Avison (Wood-Harper, 1985). The overall
model is given in Figure 2.3. A brief review must suffice:

The first two stages are concerned with examining a particular
problem situation and expressing it without imposing a solution. The
problem area needs to be explored as widely as possible, building up the
richest possible picture of the situation being studied (Checkland's
emphasis). The development of a Rich Picture based upon the
organisation's structures and processes is a very useful one and is
described in more detail in the next section.

Stage 3 begins at the end of problem expression and tackles the
question: What are the systems which, from the analysis stage, seem
relevant to the problem? This question has to be answered carefully and
explicitly and the resultant Root Definition will also reflect a certain
personal view of the organisation. A number of examples are given
below from Checkland to give a flavour of their formulation.

A department to employ social workers and associated staff to
build and maintain residential and other treatment facilities and to
control and develop the use of these resources so that those social
and physical needs of the deprived sections of the community which
Government statute determines or allows, to the extent to which

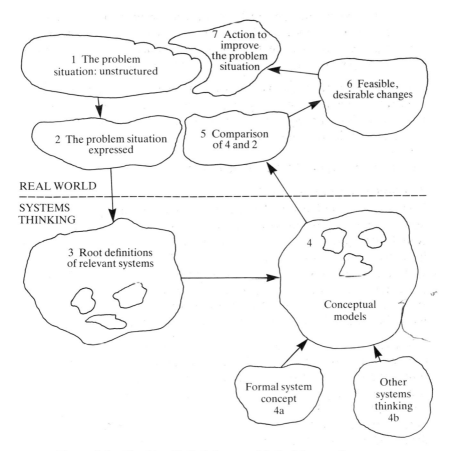

Figure 2.3   Checkland's Soft Systems Methodology: a Summary

County Council, as guided by its professional advisers, decides is appropriate, are met within the annual capital and revenue constraints imposed by the Government and Council.

(Social Services Department)

A system to celebrate a particular life style using pop-music as an emblem of the sub-culture concerned.

(A Pop Festival)

A system owned by the Manpower Services Commission and operated by the Paintmakers Association in collaboration with the

Polytechnic of the South Bank's Distance Learning Unit, to provide courses to increase technical skills and knowledge for suitably qualified and interested parties, that will be of value to the industry, whilst meeting BTEC approval in a manner that is both efficient and financially viable.

(Wood-Harper, *A Distance Learning Project at South Bank Polytechnic*, 1983)

Stage 4 makes and tests relevant conceptual models. This is concerned with what should be happening to support the requirements specified in the Root Definition: "the aim is to build an activity model of what must go on in the system". (Checkland, 1981.) Many conceptual models are in fact quite similar for different types of organisation. The example given in Figure 2.4 would be relevant to many types and sizes of enterprise.

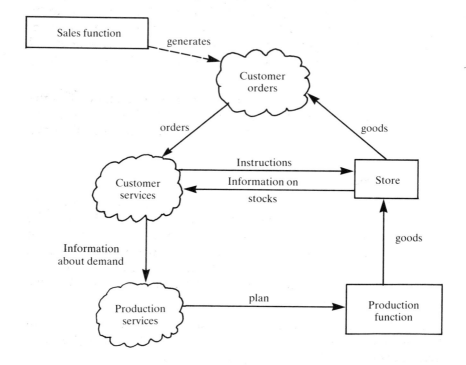

Figure 2.4   A Conceptual Model of an Order Generating-and-Processing System (from Checkland)

The next stage is to compare the conceptual model with reality. In effect this is examining the system that is required to support the Root Definition (Stage 4) with the one that actually exists (Stage 2). This will lead to the final two stages of implementing 'feasible and desirable' changes. Three kinds of change are envisaged:

Structural. Organisational groupings, reporting structures or functional responsibilities.

Procedures. Processes of reporting and informing.

Attitudes. Changes in influence, expectations and perception.

The strength of the Soft Systems approach is in its insistence upon a framework, rather than a prescription, for organisational activity. It tends to recognise the difficulty of getting to grips with a problem area and, in doing so, acknowledges a complexity which may be missing from the 'sanitised' corporate vision of the Strategic Plan. But it could be argued that its problem definition is much stronger than its problem solution. The lack of prescription, and the rather academic notions and guidelines given in the framework, may not endear it to the practising Analyst.

In the context of this book, we have chosen to concentrate upon the problem expression aspects of Checkland's work. The concept of the Rich Picture has been extended by Wood-Harper *et al*, and this forms the basis of the next section.

2.6.1   Rich Pictures

At the beginning of a systems project the people involved are likely to have only a fuzzy idea of what they want to achieve. Even if the proposed objectives can be made clear, they still need to be formulated in such a way that they can be explained to Analysts and potential suppliers. A useful technique for modelling the overall system under consideration is the Rich Picture. It attempts to show what the organisation is about.

The Rich Picture is useful to the Analyst because it provides an overall view of the current problem area as well as giving greater understanding of it. Drawing a Rich Picture should:

— Focus attention on the important issues. It is common for Analysts to become too concerned with detail too early on in a system study. Make no mistake, attention to detail is extremely important in information system design, and there are numerous techniques to support and encourage this level of precision. However, this may lead to the Analyst 'not seeing the wood for the trees', or indeed, failing to recognise that his perceived wood is actually a small part of a large forest! The use of a Rich Picture, constrained to one side of A3 or A4 paper, demands concentration on major flows and structure, stripped of their detail. In this respect it mimics the understanding of the Analyst. Users and operators are inclined to describe their tasks in detail and from this the Analyst must try to glean the overall structure and process of what is taking place. Many participants in the system are, by their very proximity to the task, unable to perceive the whole activity in which they are involved and are consequently more able to describe *what* they are doing rather than *why* they are doing it.

— It will help participants to visualise and discuss the role they have in the organisation. The Analyst has a model of the enterprise which may be incorrect, partial or biased. Inconsistencies and errors in this model are easier to sort out if the model is itself described in some pictorial form. Hence the Rich Picture can act as an important method of communication.

— It can be used to define the parts of the organisation that will be covered by the information system. The definition of a system boundary is an important task. The Analyst and the participants need to be clear about what the proposed system will cover, and – perhaps more importantly – what it will not.

— It can be used as a medium for illustrating participants' worries, conflicts and responsibilities.

There is no standard way of drawing a Rich Picture. Nevertheless, some guidelines and principles might be offered.

Begin the picture by first putting the system area under consideration into a large bubble in the centre of the page. Other symbols are sketched in to represent people, activities and physical objects of interest and importance to that system. Arrowed lines show relationships, crossed swords indicate conflict and 'think' bubbles may be used to show the main worries of the participants. The relative importance of people and things can be reflected in the size of the symbols, but this may be difficult to adopt in every instance.

Rich Pictures should be largely self-explanatory as they are designed to aid communication as well as help the Analyst visualise the problem situation. It is a method of pictorially representing three important considerations in information systems design:

— *Elements of structure in the problem area.* This might include departmental boundaries, physical or geographical layout and product types and activities.

— *Elements of process.* What takes place in the system.

— *Elements of relationships.* The relationship between structure and process is the climate of the problem area. This will include conflicts, worries and mismatches between new processes and old structures.

A Rich Picture for InfoSys is given in Figure 2.5. It also includes a 'beady eye' representing the interests of COMMUNIQUE. As stated above, there are no agreed standards for drawing Rich Pictures and you may wish to add other symbols and ideas that you feel are particularly appropriate for a certain set of circumstances. This is valuable until the picture becomes so complicated that the essential structures and processes are obscured. At this point Rich Pictures begin to have more in common with art than information system design.

The Rich Picture gives us a medium for exploring and expressing the 'problem situation' (Stages 1 and 2 of the Checkland framework). However, although it gives us a high-level view of the system, this is insufficient for information system development. Models for detailed expression of the current system are required and these are introduced in the next two chapters.

2.7 TRIGGERS OF CHANGE

Our discussion so far has concentrated upon strategic approaches to information systems development. The philosophy of top-down Strategic Planning is different to that of Checkland's Soft Systems approach. But whatever their divergence they share the perspective of 'wholeness': the perception that IS development must come from an understanding of the enterprise's whole activity and not from the piecemeal solution of problems. Indeed there are similar elements in the two approaches. Exploring the problem situation has much to do with corporate appraisal, whilst the conceptual model might be seen as part

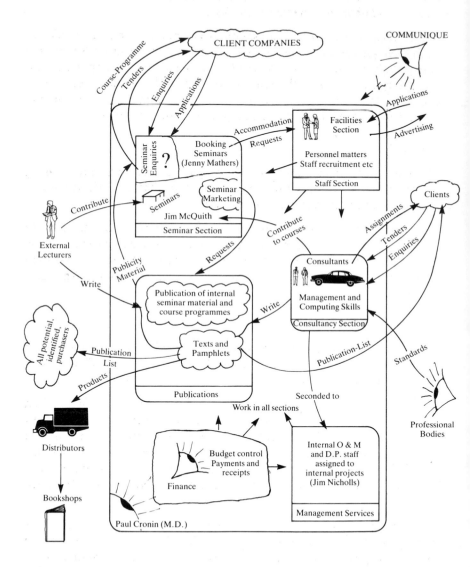

Figure 2.5   Draft Rich Picture: InfoSys

of the setting of objectives. It helps establish the desired state of the enterprise and hence will foster the formulation of objectives within the feasible and desirable changes identified.

The reasons why the organisation is considering change should be both well justified and understood. However, this trigger for change may not always be so well defined. In such instances it is important that the Analyst investigates the reasons for system development so that he can formulate the scope of the project and the climate in which he will be working. Three simple types of trigger may be identified.

Positive Trigger

"I spend a lot of my time compiling figures, producing and revising forecasts and then presenting reports and plans based on these figures. Surely the computer can help me in some way?"

In such instances the problems of the present system have been identified by those who run and use the system. The computer is seen as a possible way of getting system improvements. There is a positive commitment to computerisation – the system's user is on the Analyst's side.

Negative Trigger

"We have had problems with Stores ever since I became a Director. It's time we computerised it and got up with the times. That will sort them out down there!"

In many instances the decision to consider computerisation is taken by the system's owner, not those who actually operate the system. In such circumstances the perception of the present system's performance is unlikely to be shared by all those concerned with running the system. Indeed the operators of the system may see any investigation as a criticism of their present performance and the procedures that they have created.

In the example represented in the above quote, the company's managers had 'lost control' of Stores. They had been looking for an excuse to impose new controls for some time and computerisation presented that opportunity. All the members of the Stores section resisted the plans. They felt that management did not understand what happened in the Stores section and staff morale was low. They felt aggrieved that the systems that had been developed over many years were now deemed to be inadequate by people who, they believed, did not understand them.

Both sides had valid points, but pity the poor Analyst stuck in the middle! In such projects the chance of failure is much greater because many of the key personnel in the system are actually willing that failure. In these circumstances the climate of implementation, the time-scales and the level of ambition are all very different to projects undertaken in response to positive triggers. The Analyst has to tread more carefully, take more time and be less ambitious in the proposed changes.

Such projects are very common. The problems they present are too fundamental to be solved by rational argument or advisory meetings and presentations. The Analyst has no option but to get on with the system under conditions and contraints that may, at some time, seem to be intolerable. In such instances the computer is being used as a substitute for management and the Analyst must recognise this and plan accordingly.

Neutral Triggers

"The insistence of the auditors that we improve our controls and reporting systems finally convinced us of our need to computerise."

Neutral triggers are events that are outside the scope and control of the system. These may be as a result of Government instructions (a need to present accounts in a certain way), legal requirements or the advice of external agencies such as professional associations or auditors. These triggers have little positive or negative effect on the project in hand.

The reasons behind making the large, and sometimes traumatic, step to computerisation undoubtedly affect the rest of the project. Thus it is necessary to probe and understand these reasons to help gauge the climate of the project and to adjust time-scales and objectives accordingly.

Fortunately, an approach to project selection that recognises the strategic importance of information systems should reduce the time needed to dwell on understanding the system triggers. The justification for the project has been discussed and debated at an earlier stage and at a more senior level, and so by the time it comes to implementation the need for the project should be well documented and understood. This is not to say that all problems will be avoided, but many of the details of principle will have been resolved. Participants should be aware of the *why*, *how* and *when* of the project and its relationship to the rest of the enterprise's plans and activities.

2.8 SUMMARY

This chapter has reviewed some of the issues that will be encountered early in a systems project. In summary it:

— Suggested that a computer system can help the doers and makers of an organisation obtain benefits but that it cannot obtain such benefits in itself. This led to a consideration of how information systems can be viewed in the perspective of business activity and objectives.

— Illustrated a 'top-down' approach showing how very broad objectives can be sharpened into specific goals that have information system consequences. It was suggested that these goals should be described in the context of the business (ie not have computer and non-computer contributions) and that they should be qualified and time constrained.

— Recognised the clinical precision of strategic planning and introduced a complementary approach which acknowledged the rather fuzzy nature of organisations, problems and problem solving. The concept of a Rich Picture within the Soft Systems methodology was given particular emphasis, and an example was drawn for the InfoSys case study.

— Acknowledged that triggers of change are usually explicit in companies that have a strategic approach to planning and information system development. This is not necessarily true for enterprises that practise a more piecemeal development. In such instances it is important for the Analyst to probe the reasons behind the desire for change and to recognise the implications of these reasons. The ambition, scope, approach and time-scale of the project can then be judged accordingly.

2.9 PRACTICAL EXERCISES AND DISCUSSION POINTS

Wonderbras is an established medium sized company based in the Midlands manufacturing ladies' lingerie. It employs a small sales force, supported by a number of agents, who operate within specific geographical areas of the United Kingdom. A typical area will cover between 120 and 130 customers, of which approximately 30% will account for 80% of the business. The customers will vary from small private shops to wholesalers and department stores. Those customers held as key accounts will be regarded as priority customers, maintaining a greater level of business and requiring a higher level of service and support.

Wonderbras offers a set of permanent fashion items, but during the course of the year other products are added and deleted to reflect seasonal requirements. The range is completely reviewed on a six-monthly basis as the business is governed by changes in demand and fashion.

In the Field

The sales representatives are ultimately responsible to the Sales Director operating from the London showroom. Their performance is monitored on a monthly basis, comparing sales targets against a computer generated commission report. Two of the representatives also act as Regional Managers (North and South) to whom the other representatives report weekly by telephone. The Regional Managers use the provided information as a basis for a marketing report which is submitted to the Sales Director.

The reporting mechanism of the sales representatives is ill-defined and there is a general lack of discipline and control. Also, there is a reluctance to delegate responsibilities and decision-making powers to representatives in the field and at the point of sale.

The sales representatives are responsible for planning their own itinerary of work. Every week they compile a journey plan that will account for 90% of their visits in the coming week. It is likely that ad hoc calls will also be made to key customers. The representatives will be involved in selling, merchandising and undertaking stock checks in order to make 'bookings'. The latter activity involves physical counts and recording 'repeat orders' on the stock card. This information is copied onto a separate order form and returned to head office for processing.

The amount of stock a customer will hold at any one time is subject to negotiation and reviewed periodically. However, with key account customers, a Regional Manager or the Sales Director will agree stocking arrangements. The representative will then plan a 'product mix' for key account customers that corresponds to the agreed merchandise value negotiated by the management staff.

Bookings are confirmed as orders by the return of a copy of an order from head office to the representative. The only other contact is the submission of the representative's journey plan, weekly report and expenses. In return the representative receives a monthly commission and an aged debtors report.

The paperwork procedures are regarded as cumbersome, repetitive and sometimes unnecessary. The representatives receive no analyses such as order summaries with breakdowns by product or value. There is no mechanism to indicate when a booking has been converted into an order and has been subsequently fulfilled. A customer's stock record card may indicate requirements outstanding over long periods, but these may be the result of Wonderbras's poor stocking responses and not a reflection of market trends. Inadequate operational procedures can generate artificial demands and impose extra strain on the company. Representatives will occasionally be involved in special promotional drives, although the level of support makes such activities unattractive and cumbersome to administer.

It is evident that the sales representatives are experiencing problems due to the nature of the structure and processes (or lack of them) of normal operation. The customer rarely receives delivery of all the standard items ordered because of stock problems at the warehouse. In such cases the customer may cancel the remainder of the order and the representative will then be involved in making repeated, unplanned calls to the customer outside his usual schedules. Wonderbras's failure to respond rapidly often means products are available later than required by the representative, consequently building up a redundant stock that is difficult to sell.

There are products in demand that the company cannot or will not produce and distribute. During the course of a season (winter or summer) the representatives face competition from other major manufacturers with a product range that reflects seasonal demand and fashion changes. Each season the introduction of new products causes problems. The representatives are unable to provide adequate support to customers with portfolios, specimen material and comprehensive illustrations. Although the company will test a product in the market and assess it on criteria of advance bookings, it is difficult to measure response because of stock problems. Moreover, once a decision is made to introduce a new product the company is committed to it regardless of market demands.

The level of support provided during the product life cycle is not planned or measured. There are no national campaigns or themes, and promotional aids and display materials are scarce; the company relies heavily upon a product reputation for 'quality'. The representatives experience difficulties in attempting to operate in an ill-structured framework. They do not feel a valued part of the organisation owing to lack of formal meetings and absence of support information (stock availability, deliveries, performance, etc).

The sales staff feel that there is no coordinated management of the entire selling effort. This is shown by the little feed-forward information made available regarding products, territories and markets. In addition they have difficulty in identifying with the organisation; the policies are not translated into plans, only personal targets. The potential role of the representative remains unrecognised. The representatives operate in the field and so are able to gain information about the changing market-place, such as the role and attitudes of customers, consumers and competitors. Valuable feedback information could aid decision-making procedures and prevent the phasing out of successful products.

Back at the Sales Office

Wonderbras employs a small sales administration unit of four clerks (three full time and one part time) responsible to the Sales Administration Manager based at head office. The sales office personnel are responsible for handling all customer enquiries and processing orders received by the company. The orders can originate from the customer direct or from the sales representatives. The latter account for 75% of the total orders received.

The Sales Administration Manager is responsible to the Sales Director and controls the daily operations of the sales office. Decision making is limited to allocating the priority of customer orders. The sales office operates both computer and manual systems during the processing of orders. The clerks in the sales office enter the orders and the computer room operators are responsible for organising despatch and invoicing. Information generated by the system is used for analyses and reports which the office distributes on a regular basis. The sales office also maintains informal contacts with the representatives in the field and provides information on request.

The sales office handles all customer enquiries, which are usually about the progress of orders, although a large proportion are also complaints about deliveries. The computerised stock allocation system allocates items to customer orders according to the priority level of the customer and the availability of stock. In the case of out-of-stock items, the system will perform one of two alternatives – cancel the whole of the order or accept the part-filled order. The larger customers (approximately 40%) cancel remaining items and reorder, whilst smaller customers are willing to accept part-filled orders. When new stock becomes available the system allocates items to the oldest order on file. The sales office is also responsible for preparing and ensuring that all export and sundry documents are made available for orders. The despatch, invoicing, credit control and after-sales functions are distributed amongst other departments.

The Sales Administration Manager compiles two reports. The New Range Analysis is prepared during the early part of a season and based on advanced bookings submitted by representatives. Call analyses are prepared quarterly to report on salesman performance, based on calls, repeat orders, value of orders and new orders.

The sales staff are currently running two systems, the new computerised one and the old manual one. The sales office has found it impossible to dispose of the old system because of perceived inadequacies of the new one. The stock allocation system has been devalued by the problems of stock availability. The order staff have had to develop additional manual systems to process orders prior to entry to override customer priority. The new system cannot respond to demands made by staff. The computer is mainly used for data entry and the software gives very limited interrogation facilities. Nearly every order (average 50–60 items) requires manual reference to print-outs. An estimated 80% of orders cannot be fulfilled on receipt, hence the sales office is able to achieve only a 20% service level.

Some of the problems emanate from the structure and dual policies currently practised. Wonderbras operates with both centralised and decentralised processing, resulting in repetitive and unnecessary activities and excessive paperwork transfers. Responsibilities for orders are spread over the company structure, preventing order staff from monitoring, controlling and providing information on order status. The system designed and implemented to date resulted from two consultative meetings with sales staff. The intention is that the system will eventually provide a comprehensive reporting system.

Directing it All

The Sales Director is responsible for all marketing and sales operations within the United Kingdom. Based in London, with weekly calls to the Midlands Sales Office, he controls and directs the total selling effort of Wonderbras. Marketing is limited owing to the absence of any dedicated resources. In addition to his managerial duties, 50% of the key accounts are directly under the Sales Director's control.

The Sales Director monitors and controls the sales force periodically with the aid of commission reports and call analysis reports prepared by the sales office. The Regional Managers submit a brief weekly marketing report, but the information is too general to permit comprehensive analysis. Contact with the sales representatives is mainly informal.

Wonderbras maintains regular contact with an agency that provides specific marketing services. These include arranging photographic sessions, presentations, media bookings and limited marketing studies and information. The Sales Director's growth strategy is concerned with increasing sales of existing products and developing and introducing new products for current market areas.

There is little investment in the media, as the market is volatile to seasonal changes and advertisements do not significantly affect the behaviour of the consumer, which is reliant on impulse 'point-of-sale' purchasing. The formulated strategy is to achieve maximum continuing sales over more than one season.

The Sales Director prepares a forecast covering expected sales for the coming season. The forecast covers the whole product range and is based upon historical demands measured by actual sales and bookings. The Sales Director also applies a level of confidence of a product's expected performance in the market. The forecast is the mechanism that triggers production. It highlights any variance (shortfalls) between actual and required stock and between sales and bookings. The current frequency of forecast monitoring is limited to monthly.

The other main function of the Sales Director is the introduction of new products to supersede old ones. The initiative to undertake this stems from the seasonal fashions, ideas and the level of confidence in a new product/colour. These factors are then translated into a product brief which is given to the designer. A new range analysis, that records advanced bookings, gives a measure of customer response to a new product two weeks after its launch. The critical time limit is set at six weeks for any strategic decisions concerning a new product. Other tasks personally executed by the Sales Director include campaign planning (twice yearly) which is based on resources set in the budget. The Director also negotiates with the central buyers of four key accounts to sell Wonderbras products under their own brand names.

The response by the production department is poor and so the stock variance can be unacceptably large. Consequently, Wonderbras is either out of stock or late availability means redundant stock has to be sold off through the trade. The ability to monitor and control the forecast is negligible. The monitoring is done only monthly and it takes three months for control action to take effect. This is due to the long lead times of the manufacturing cycle. Wonderbras has a poor feel for its markets and is unable to measure reliably its share and determine its market potential.

It is evident that a mismatch exists between strategies and objectives of Wonderbras and its ability to achieve them. It cannot respond to demands made by such a dynamic environment and hence its very survival is threatened.

Tasks

1 Draw a Rich Picture which summarises Wonderbras.

2 What are the primary tasks of the company?

3 Make a list of the issues that reflect the different perspectives expressed in the Rich Picture (those topics or matters of concern that are subject to dispute in pursuance of the primary tasks).

4 Suggest several possible root definitions of relevant systems. Agree on a composite root definition on the basis of negotiation.

5 Draw a conceptual model of the system based on the agreed root
   definition (this should show the activities that the system must
   undertake in order to be the system named in the root definition).

3 The Feasibility Study

3.1 INTRODUCTION

Once a preliminary area of application has been established, it may then be subjected to a more rigorous examination in a Feasibility Study. The Analyst will, of course, already have formed some ideas about the application from the preliminary planning work. However, the Feasibility Study represents an opportunity to 'firm up' knowledge of the system and to form ideas about the scope and costs of possible solutions. In many respects the Feasibility Study is a quick and dirty mini-systems analysis with the Analyst being concerned with many of the issues and using many of the techniques required in later detailed work. Feasibility Studies are usually undertaken within tight time constraints and normally culminate in a written and oral Feasibility Report. The content and recommendations of such a study will be used as a basis for deciding whether to proceed, postpone or cancel the project. Thus, since the Feasibility Study may lead to the commitment of large resources, it is important that it is conducted competently and that no fundamental errors of judgement are made. This is easier said than done.

This chapter examines three different aspects of feasibility and the compromise that eventually has to be made between them. The results of this compromise are then presented in a Feasibility Report which recommends whether a project should progress to the detailed investigation stage.

3.2 THREE TYPES OF FEASIBILITY

In the conduct of the Feasibility Study the Analyst will usually consider three distinct, but interrelated types of feasibility.

3.2.1 Technical Feasibility

Incorporate in to costing tool

This is concerned with specifying equipment and software that will successfully support the tasks required. The technical needs of systems will vary considerably, but might include:

— The facility to produce outputs in a given time scale, eg 20,000 examination certificates in three weeks.

— The ability to provide certain response times under certain conditions, eg no more than a two-second response time at each terminal when there are four terminals being used simultaneously.

— The facility to input a large number of documents in a limited time scale, eg 400,000 gas readings in one day.

— The ability to process a certain volume of transactions at a certain speed, eg to report on seat availability and record airline reservations without a significant delay to the passenger.

— The facility to communicate data to distant locations, eg regional sales figures transmitted to an American head office.

In examining technical feasibility it is the configuration of the system that is initially more important than the actual make of hardware. The configuration should show the system's requirements: how many workstations are provided, how these units should be able to operate and communicate, what input and output speeds should be achieved and at what print or screen quality. This can be used as a basis for the tender document against which dealers and manufacturers can later make their equipment bids. Specific hardware and software products can then be evaluated in the context of the logical needs.

At the feasibility stage it is possible that two or three different configurations will be pursued that satisfy the key technical requirements but which represent different levels of ambition and cost. Investigation of these technical alternatives can be aided by approaching a range of suppliers for preliminary discussions. The technical performance and costs of rejected alternatives should be documented in the Feasibility Report.

3.2.2 Operational Feasibility

Operational feasibility is concerned with human, organisational and political aspects. General impressions of these factors may be gained from the Corporate Appraisal and through consideration of the system trigger. Amongst the issues examined are:

— What job changes will the system bring? Most people react unfavourably to change. Planned job changes must be carefully handled so that those affected are seen to gain in a way that they feel is acceptable. This may be through job enrichment or simply through raising wages.

— What organisational structures are disturbed? The suggested system may cut across accepted organisational relationships and threaten the status of individuals and their promotional expectations.

— What new skills will be required? Do the current employees possess these skills? If not, can they learn them? How long will they take to learn?

It is unlikely that a project will be rejected solely on the grounds of operational infeasibility but such considerations are likely to critically affect the nature and scope of the eventual recommendations.

It should also be recognised that although the computer brings significant alterations to an organisation it is not the only source of change. The way that an organisation manages and implements changes in other areas should be considered when computer related changes are planned. Agreed procedures for discussing proposals, Staff Consultative Committee, Trade Union Agreements, Staff Forums, etc, should be identified and followed.

3.2.3 Economic Feasibility

Many organisations evaluate projects on an economic basis – they must show financial returns that outweigh the costs. For this reason, management tend to give more weight to economic feasibility than to technical and operational considerations. A number of approaches to assessing the costs of solutions have been suggested. Approaches include:

Least Cost. This is based on the observation that costs are easier to control and identify than revenues. Thus it assumes that there is no income change caused by the implementation of a new system. In such an evaluation only the costs are listed and the option with the lowest cost is selected.

Time to Payback. The 'Time to Payback' method of economic evaluation is an attempt to answer the question: "How long will it be until we get our money back on this investment in systems?". This requires data on both costs and benefits.

In the 'Time to Payback' method, the alternative which repays the initial investment the quickest is selected. This method of evaluation has two significant disadvantages.

— It only considers the time taken to return the original investment and ignores the system's long term profitability. Thus options that are more profitable in the long run are not selected.

— The method does not recognise the time value of money. Benefits that accrue in the distant future are not worth as much as similar benefits that occur more quickly, but the 'Time to Payback' method fails to recognise this.

Net Present Value. This is a well defined and practised method of economic evaluation. It builds in an allowance for the 'time' value of money, represented by the Present Value Factor. In this method the net benefits are reduced in value by applying this factor, so reducing the value of a benefit to its present worth. For example, benefits of £30,000 planned for 10 years' time will actually be worth only £4800 if a Present Value Factor of 20% is used. Thus benefits that appear late in the project's life span contribute little to its economic feasibility. This may make project selection a little conservative. In many computer based projects, benefits appear late in the project life and heavy discounting of these benefits may lead to the selection of less ambitious projects that yield a quicker return.

Breakeven Analysis. This technique is particularly useful when the system is subject to varying workloads. It distinguishes between fixed and variable costs and fixed and variable benefits. The data is plotted on a graph where the vertical axis is the amount of cost or benefit, and the horizontal axis is the increasing level of the workload. The fixed costs are plotted first with the variable costs plotted above them to show the increase in total costs as the workload increases. The same is done for the benefits. The crossing point of the total benefits line with the total costs line indicates the breakeven point. Workloads to the left of this point do not justify the use of the system, while workloads to the right of the breakeven point do. Breakeven analysis does not give the full picture needed for the economic evaluation of systems, but its emphasis on the operational phases can be useful.

There are a variety of methods which may be used in economic evaluation. These methods may give contradictory advice and none of them enjoys universal acceptance. The financial evaluation of alternatives is a common requirement in business, not just in information systems development. You may wish to consider this aspect more closely by examining a paper that looks at economic feasibility of computer systems in detail (Chapin, 1981) and a book that gives an in-depth look at the whole topic (Lumby, 1981).

However, whichever economic evaluation method is adopted (except the rather simplistic Least Cost) there will be a need to predict and quantify benefits. This is typically more difficult than quantifying costs. Chapin identifies several reasons for this:

— Uncertainty about the timing and amount.

— Problems of expressing certain benefits in direct monetary terms. What is the value of not having to apologise for as many order errors? Many benefits will often appear as 'intangibles' – better management information, improved management controls, etc.

— The benefit is often due to a joint effort of a number of departments in the organisation. In such cases it is very difficult to assess the computer's contribution.

You may recall that this latter issue was explicitly recognised by suggesting that all costs and benefits incurred in a system project should be included, not just the computer system's contribution. This is a further justification for taking a 'top-down' approach to project selection through seeking business objectives. The advantage lies not only in the inclusion of all costs but – more importantly – the benefit itself is defined by the goal. The significance of this must not be missed. Many conventional projects are subject to a set of costs that must be offset by scrabbling around for presumed benefits. These costs are often set in an inappropriate way – someone 'decides' that the budget will be £10,000, because this 'sounds a reasonable figure'. All subsequent costs and benefits have to be squeezed into this framework. Benefits may be hard to find, so intangible gains such as better control, up-to-date reports, etc are introduced to make the project look financially attractive.

But if the alternative perspective of defining benefits before costs is adopted, then the framework for system development immediately becomes more realistic. For example, the overall ambition of the project should become clearer. The budget for computer hardware and software can be set within the benefits that are likely to accrue. If savings of £250,000 per annum are envisaged then it is probably unrealistic to restrict thinking to £10,000 systems. A recognition of this will permit more innovative technical solutions and operational strategies. One of the authors of this book recently worked on a project that was blighted by the Managing Director's insistence of an £8,000 budget. It was not possible to develop the system for this money and the project was subsequently dropped. It was later pointed out that development at this price would permit the system to 'Payback' in six weeks, a performance not matched (or required) by any of the company's other investments. The Managing Director acknowledged this but was unmoved. A consultant had told him that a system could be bought for the price of an average saloon car and he had no reason to doubt that advice.

3.3 INFOSYS: A FEASIBILITY EXAMPLE

One of the goals identified earlier was to increase seminar income by £300,000 per annum within three years. The Seminar Manager, Jim McQuith, felt that this target could be achieved given the appointment of additional staff and the improvement and extension of the information system used to support those staff. Preliminary discussions identified the following feasibility issues:

Technical. It was pointed out that seminar staff consider form-filling to be unproductive. An increase in clerical work caused by the need for more information would be unacceptable. Thus data capture should be as discreet as possible.

It was generally felt that the information system should be available to all staff in the seminar section and so a multiuser system was essential. "We don't want to be juggling floppy disks," said McQuith.

Input transactions were considered to be fairly straightforward (although subject to some operational constraints) and most outputs were not time-critical. Possible exceptions to this were thought to be current seminar bookings and seminar content.

McQuith, while accepting the need for better information system support, did express some concern:

I see little point in offering such facilities if we are going to be hanging around for 15 minutes awaiting an answer. Furthermore, the system must be completely reliable – both in the information it holds (the seminar figures have just got to be correct) and its functioning. We don't want any of these 'Sorry the computer is down' messages. We are going to be completely reliant on this.

Another aspect of reliability was mentioned in the preliminary discussions held with Paul Cronin:

You must remember that this system has to be 100% secure. It holds valuable customer data. I shudder to think what would happen if our competitors got their hands on it.

Operational. The staff interviewed were generally very positive about the project. They understood the reason for its development and indeed, in many instances, had contributed to the planning meetings that had instigated it. Two significant worries could be identified.

The first was a concern that the computer system would not damage their current activity. "We want it to help and guide us, not to impose constraints, problems and extra workload."

There was also a feeling that the system's use as a monitoring device could outweigh its support facilities.

I'm very worried. It seems that it will give McQuith a tremendous amount of information about our activities. I can imagine him asking all sorts of trivial questions about our work. We could get into a Big Brother situation and that would cause a lot of problems.

Secondly, seminar staff were very worried about the actual operation of any computer system. Simplicity was stressed:

We're computer simpletons and we want to remain that way. We want to know what buttons to press and what to do and who to contact if it all goes wrong. Don't try and turn us into computer whizz-kids.

Training was briefly discussed and concern was expressed about the time that this would take.

Economic. It is difficult to do an in-depth financial analysis at this stage. The actual form of the project is still largely unknown, as is the scope of other possible applications. A computer may be purchased solely for this task or it may be just one of the jobs of a much larger computer. No decision has yet been made on the various methods of procurement – buy, lease or rent – and this will also affect cash flow. However, a few 'ballpark' figures can be produced to gain an impression of the amount of investment likely to be available for the project. Starting with the benefits:

		£
Benefits:	Year One	100,000
	Year Two	225,000
	Year Three	300,000
	Year Four	300,000
	Year Five	300,000
	Total	1,225,000

It is probably appropriate to use a fairly blunt evaluation principle at this stage. COMMUNIQUE recommends that all computer-related projects should payback in three years. Using the principle:

$$\text{Time to Payback} = \text{Investment/Annual Average Benefit}$$

If the project life is defined as five years:

$$3.0 = \text{Investment}/245,000$$

Hence: Investment = £735,000

This figure gives a reasonable feel for the size and scope of the project. If our investigations show that such an investment is too low then we either have to accept a higher investment, reassess the technical options, reconsider the operational constraints or abort the project and look for new ways of achieving the defined goals.

We will clearly have to do a more detailed costing study in the future but this may only be possible when the overall IS strategy becomes clearer and when the detail of this particular project emerges from our subsequent fact-finding.

3.4   THE FEASIBILITY REPORT

The Feasibility Study will usually culminate in a formal written report and an oral presentation. The possible contents of a feasibility report,

based on suggestions of Collins and Blay (Collins and Blay, 1982), are
listed below.

Introduction

Background to the project. A brief review of the layout of the presented
report.

Terms of reference. This is likely to include reference back to the
preliminary analysis and an explanation of how the system under
discussion was selected as a candidate for investigation. It will also
include details about the scope, resources and time-scale of the study. It
is important that these are established and agreed at the outset. Possible
terms of reference for the sales support project might be:

— To undertake a Feasibility Study of the sales support function
   identified in the Corporate Strategic Plan (see Section 2.5). The
   study will commence on 1 August 1987 and be undertaken by one
   Senior and two Junior Analysts. A formal report will be presented
   to members of the Information Systems Planning Group on 11
   November 1987, followed by an oral presentation at the Group
   meeting on 24 November. The Feasibility Report will adhere to
   company standards (see ref 45/23/65) and evaluation criteria.

Terms of reference should be explicit about what is expected, by when
and what resources are available to achieve this. The intended outcome
from this project stage is particularly important. Many projects are
blighted by unfulfilled expectations.

"We expected to see a system not a report!"

Parkin (Parkin, 1980) has used the term 'deliverables' to describe these
outcomes. Terms of reference should always include the intended
deliverables. In the example given above, these are a report and a
presentation. In other instances they might be a system, a program, a
memorandum, etc.

Existing System. A description of the relevant system(s) currently
operating in the organisation. These will have been investigated using the
fact-finding techniques described in the next chapter and presented using
appropriate methods such as system flowcharts and data flow diagrams.
These will be less detailed than in the subsequent detailed analysis, but
special attention must be paid to the technical requirements. Particular
problems will be highlighted and the implications of these discussed.

System Requirements. These will be derived from the existing system (outputs currently produced may still be required when the system is replaced) and from discussion with system users and operators who have identified requirements that are not presently fulfilled. Critical performance factors must also be covered, (eg, the need to produce 5000 invoices per day, to process transactions in less than 5 seconds) because these will have an important bearing on the hardware selection. Audit, Security and Data Protection implications may also be discussed.

Proposed Systems. An outline logical system design may be presented, together with sketchy definitions of inputs and outputs. These will be described more in their content than in their layout and display.

The differences and advantages of the proposed system over its predecessor will be highlighted, together with its effect on other systems currently operating in the organisation. The new system may impose certain constraints in operation (eg, all input documents must be submitted by 4 pm) and these should be clearly described and discussed.

The possible effects on staff must be identified and a strategy for staff training, reduction or redeployment suggested or requested.

The equipment and software requirements of the system might also be described. The extent to which this can be done depends upon the current resources of the organisation. If the firm already has a large computer then the extra hardware is likely to be additional terminals, more secondary storage, perhaps more memory. It is also likely that these will not be required until the system becomes operational and so the specification may be altered as detailed analysis and development clarifies the nature of the system.

In contrast, large projects and organisations without significant computing resources will have to invest in hardware before development can get under way. As a result there are significant pressures to select and purchase the hardware very early in the project's life, before many of the detailed implications of the system have been discovered.

Plans will also have to be made for security and disaster recovery. For example, the purchase of two machines may be considered so that if one fails the other can be brought swiftly into operation.

Development Plan. Suggested project plan and policies for phasing in the project and managing the transition from the present to the proposed system.

Costs and Benefits. These have already been discussed. They will clearly vary in detail and accuracy, as will the techniques used to evaluate them.

Alternatives Considered. In the process of arriving at a suggested system the Analyst usually considers and rejects a number of options. It is important to record these considerations for two main reasons. Firstly, it may nip a number of time-consuming "have you considered . . ." discussions in the bud. Secondly, it permits the sponsor of the study to examine the legitimacy of the reasons for rejection. For example, the Analyst may have rejected a certain option because, in his impressions gained from the preliminary analysis, it appeared to be too costly. However, the information contained in the study may now persuade the sponsor to change his mind about the level of ambition of the project and so the rejected alternative becomes feasible. This alteration would be unlikely if details of rejected alternatives were not included.

The report would normally end with Conclusions and Recommendations and relevant Appendices.

3.5   THE FEASIBILITY COMPROMISE

The three ways of approaching feasibility are likely to conflict. In general, 'better' technical solutions cost more money, while robust, helpful, user-friendly software is time consuming to write and therefore incurs high development costs. Such software may also mean larger programs and, because the system has to carry a much larger software overhead, this may begin to conflict with performance requirements. In many instances technical and economic factors become paramount – "the system must have a two-second response time and return its investment in three years" – and so the operational factors become devalued. This often has unfortunate consequences.

The Feasibility Study differs from analysis 'proper' in its level of detail. It is difficult to give general advice on what constitutes an acceptable depth of analysis because this will vary with the organisation and the application. There is always the nagging worry that the detailed analysis work will uncover a hitherto overlooked fact that now makes the project infeasible. This is further complicated by the difficulty of reconciling the three feasibility criteria, particularly with the insistence of many organisations on an economic cost benefit analysis. The restricted time-scale of a Feasibility Study also makes it difficult to comprehensively evaluate and offer sufficient options at different levels of cost and ambition.

For microcomputer applications there are special difficulties in conducting Feasibility Studies. In large organisations the task may be given to a Senior Analyst, but in a firm computerising for the first time there is no equivalent person. Thus the enterprise is very dependent upon its own non-specialist staff and the integrity of possible suppliers. In most instances the suppliers of microcomputers will not have the necessary resources, skill or time to perform a proper Feasibility Study. It is difficult to justify even one day of analysis on a job with a likely profit margin of less than £1000. Thus the preparation for computerisation may be less than ideal.

Prototyping (See Chapter 1) may have an important role to play in these early stages of the project. Two examples must suffice:

— In one instance the Analyst was concerned about meeting the technical output requirements identified in preliminary discussions with the main user. If these could not be attained using relatively simple technology then the cost of purchasing and maintaining a new, advanced printer would render the project economically infeasible. This problem was resolved by creating a set of test programs that produced and timed representative output. The quality and speed of the printing was checked with the user who agreed that they met his requirements. This impressed the user as well as eliminating one of the Analyst's doubts.

— In another project the Analyst was faced by a set of operators who had already suffered from a poorly planned computer installation. They seemed sceptical of his plans until he used a commercially available package to demonstrate the opportunities that existed. The operators made detailed criticisms, but their attitude towards the whole project became more positive as the session progressed. The simple idea changed aspects of the operational feasibility within one hour!

3.6 SUMMARY

Once possible areas of application have been identified they should be subjected to a Feasibility Study. This chapter has:

— Identified and described three feasibility criteria – technical, operational and economic.

— Introduced four approaches to economic feasibility and demonstrated one of the options in the context of the case study.

— Outlined the contents of a Feasibility Report.

— Recognised the compromise between the three feasibility criteria and the possible role of Prototyping in arriving at that compromise.

The Feasibility Study will culminate in the Feasibility Report which will be presented to management for approval. If this approval is forthcoming then detailed analysis and design will commence.

3.7 PRACTICAL EXERCISES AND DISCUSSION POINTS

Spareparts plc already has a computer system in operation in various parts of the company. Six months ago Mark Jenkins, the Sales Manager, requested a Feasibility Study to assess the practicality of automating his Sales Analysis system. The present system is run by two sales clerks who receive copies of the orders taken and enter them in statistics books. They produce the following reports quarterly.

Sales by Salesman, showing the following details for each salesman:

— The value of each of the six types of part he has sold in the preceding three months, compared with the values for the corresponding three months of the previous year.

— The total value of sales in the corresponding periods.

— The salesman's commission in the corresponding periods.

Sales by Customer within Area, showing:

— For each customer, the value of each of the six types of part which has been purchased in the current quarter and the corresponding quarter in the previous year, together with the totals for the other quarters.

— The total value of each type of part sold in each area of the company's operations, together with the total for the company as a whole. Again, comparison figures are given for the corresponding quarter in the previous year.

There are more than 70 salesmen and over 4000 customers in 120 different areas. The present system has two main problems:

— The totals of the two sets of reports are frequently in disagreement.

— The reports are not ready until eight weeks after the close of the quarter.

The Sales Manager is interested only in knowing of downward trends and cases where an upward trend exceeds 15%. The majority of figures show small upward trends, and he has to wade through all these to find the few cases he is interested in.

The Feasibility Study has now been completed (the cost of the study is to be written off as an overhead). The financial findings of the study are:

Direct cost of the present system	£3500 pa
Indirect cost of the present system	£1500 pa
Cost of computerisation project:	
Systems and programming costs	£6000
Additional temporary staff (needed to code	
customer and area information)	£2000
Data entry costs	£1000
Computer time to convert and test the system	£1000
Extra data collection and coding costs	£ 500 pa
Operational costs of new system	£1500 pa
Cost of maintaining the new system in the first year	£2000

The Sales Manager estimates the benefits to be valued as follows:

Increased accuracy	£200 pa
Increased timeliness (one week instead of eight)	£600 pa
Savings in time by computer highlighting exceptional cases	£200 pa

The clerks are to be reassigned to other activities as order data is already entered into the computer elsewhere. The life of the system is estimated as five years. Unfortunately, in years four and five the system would be displacing other planned systems (a negative benefit of £3000 is to be assumed for each of those years).

Tasks

1   The following questions relate to the Spareparts case study above:

(a) What is the payback period of the proposed system?
(b) What technical features must the new system achieve?

(c) What operational constraints are likely to apply to Spareparts PLC?

(d) Overall, do you consider the computerisation of the system to be feasible and desirable? Justify your answer.

2   There are two main approaches to project appraisal using discounted cash flow. These are Net Present Value (NPV) and Internal Rate of Return (IRR). Investigate these approaches and present a short tutorial paper that shows the use of these two methods in a hypothetical example.

4 Fact-Gathering Techniques

4.1 INTRODUCTION

Analysis is primarily concerned with three tasks:

1 Finding facts that will permit understanding of the present system and aid the design of any successor.

2 Mastering fact-finding techniques that enable the finding of these facts.

3 Organising the facts into a rigorous set of documentation.

This chapter concentrates on the middle issue: fact finding techniques. But before embarking on an examination of these fact-gathering techniques, it is useful to discuss briefly the other two aspects of this trio so that the context of this chapter can be better understood.

In many respects the whole of this book is about what facts to find and how to record these facts. The models that are constructed in this and subsequent chapters demand information that must be uncovered. Thus, for example, the Corporate Appraisal (see Chapter 2) requires facts about the economic structure of the industry, the competitive position of the company and the climate of industrial relations. The System Flowchart (Chapter 5) will demand identifying administrative arrangements, information flows between sections and the location of files. Data flow diagrams and their supporting data dictionary entries (Chapter 6) will need logical processes which have to be specified in detail. The same investigative background is true of Entity-Relationship models (Chapter 7), their very construction will raise problems that have to be resolved through subsequent fact finding. Two points must be stressed:

— Facts will not emerge in neat bundles. General issues about the department are likely to be wrapped up in administrative detail and discussions of office politics.

— It is impossible ever to be sure that fact finding is 'complete'. Partitioning of analysis and design is likely to be counter-productive (see Chapter 1) since it is an iterative activity.

It is the fact-recording techniques that prompt what facts to find and give the protocols for modelling them. Thus documentation standards attempt to direct fact recording by requiring the completion of a standard set of diagrams and documents. The role of these standards is introduced in this chapter and examples of their use are found throughout the book. For example, a data flow diagram is constructed using a standard set of symbols and rules which are introduced when the construction and purpose of this model is considered.

Documentation standards attempt to be comprehensive, but the very variety of information systems means that any attempt to set a standard for all fact collection would be hopelessly bureaucratic (see Chapter 1). In practice, most organisations adopt their own 'house standards' and documenting conventions, which they have found over time to suit their needs.

4.2   FACT-GATHERING TECHNIQUES

4.2.1   Background Reading

Organisations usually amass a considerable amount of documentary evidence and this can provide the Analyst with an important insight into current organisational norms. This evidence may be available in many forms, although it must be recognised that smaller companies may be less formalised or documented because of time and resource pressures. Reports of previous market surveys or Feasibility Studies may be outdated but worth reading to understand the background of the current study. Company information may be available in the form of:

— organisation charts;

— administrative procedure manuals;

— job descriptions and specifications;

— training manuals and memoranda;

— sales and promotional literature.

The Strategic Plan and its associated Corporate Appraisal will also be significant sources of relevant information.

Time spent undertaking this type of general fact finding will depend upon circumstances. There can be a tendency to collect too much information because the Analyst is in relatively trouble-free waters. Some commentators (Yourdon, 1986) have specifically discouraged the practice of modelling the existing systems, because "systems people tended to spend far too much time on this phase of the project". Some call it 'analysis paralysis', others (McMenamin and Palmer, 1984) – the 'current physical tarpit'. It is certainly difficult to decide when fact collection at this level has been successfully completed.

4.2.2 Interviewing

Interviews are formal meetings where the Analyst can obtain information about the operations of the present systems and the requirements of any planned replacement. Successful interviewing is a skill that can be developed through practice. Furthermore, it is undoubtedly the most common fact-finding method (see for example, Sumner and Sitek, 1986) and has to be conducted with the full cooperation of all employees. Successful interviewing is particularly important in the earlier stages of a project. Many people place a high value on first impressions, and their cooperation in the development of a new system will be hindered if the initial contact is unsatisfactory.

Interviewing will be used at various times and for different purposes as the project progresses. For example:

— to gather facts about the procedures and decisions taking place in an organisation;

— to check the Analyst's understanding of system operations with users of all levels;

— to validate aspects of a proposed system design;

— to build confidence in the design of a new information system.

The purpose of the interview will determine the balance of the discussion between the interviewer and the interviewee. In general, most of the interviews in the early parts of the project will require the Analyst to prompt, then *listen*; in the latter parts to *explain*.

Planning an Interview or Discussion

Planning is essential and may take as much time as the conduct of the actual interview. Three aspects of the proposed interview need to be planned.

 1 *The objective or purpose of the interview.* It is essential to decide what will be accomplished by the discussion. Wherever possible this should be expressed in terms of targets, such as 'to identify the store manager's view of the stock control problems' or 'to define the structure and format of the monthly stock report'. The purpose of the meeting should be clearly explained to all participants so that relevant documents and information can be collected in advance.

 2 *The time and venue of the interview.* A mutually convenient time, duration and place for the meeting must be established beforehand as this will allow the participants to schedule work accordingly. It may be preferable to be away from the normal workplace if this guarantees uninterruption and privacy.

 3 *The authorisation for the interview.* The Analyst should always obtain the appropriate manager's consent before interviewing staff. Where a series of interviews is required then it is sensible to work downwards through the organisation's hierarchy. Knowledge of the interviewee's position and duties and preparation on the subject for discussion helps generate confidence. To eliminate the possibility of arguments or differences of opinion it is wise to interview only one person at a time. Insecurity can sometimes prompt a manager to request attendance at a subordinate's discussion and the 'one-person' rule can be used to deny this. It is important to gain independent views of the system. There is often a large gap between how the manager thinks the system works and how it does actually work.

Conducting the Interview

A clear introduction and explanation of the purpose of the interview should be given. If an interviewee is left unsure of the identity and role of the interviewer then the freedom of the discussion will be restricted and the interview prove less effective.

Elementary good manners should be observed. For example, being punctual and keeping the discussion to the time agreed in advance. Conforming to certain dress requirements is a further example. There are enough problems in conducting good interviews without erecting avoidable barriers.

Many interviews do not meet their objectives simply because they are poorly controlled by the Analyst. This does not mean that the Analyst should slavishly follow a preprepared checklist and abruptly steer the interview in an uncompromising manner. Some gentle redirection of the discussion is usually enough to bring the interview back to its immediate aims. Care also has to be taken in imposing such control. Permitting the participants to develop certain themes can give very useful insights into departmental tensions and operational difficulties. This information may ultimately prove more useful than the points marked for discussion in the Analyst's preinterview checklist. Thus, a willingness to replan an interview is vital if unexpected information comes to light.

The questions asked should always be relevant to the interviewee and appropriate to his knowledge and status. It is always advisable to use the interviewee's language to describe an aspect of a system and not to introduce unnecessary jargon from the information systems world.

During the interview the Analyst should try to identify the following types of inadequate response:

— *Non-response.* The interviewee refuses to answer a question. The very act of refusal may give an important insight. An accountant's refusal to discuss his relationship with the marketing department may indicate certain interdepartmental conflict.

— *Inaccurate response.* This may occur through deliberate or accidental distortion. It is very common to find that two or three members of an organisation give slightly different descriptions of how certain operations are undertaken. It has already been recognised that there is often a distinction between how a manager thinks something is carried out and how it actually is carried out. All responses should be cross-checked.

— *Irrelevant response.* The interviewee does not give an answer to the question asked. This tends to be a waste of time but is little more than an irritation if the information can be elicited from the next, more carefully phrased, question.

— *Inappropriate question.* The interviewee lacks the necessary information to frame an adequate response. This gives some ideas about the organisational boundaries. Incorrect information about part of a certain system may suggest that the interviewee has little direct connection or involvement in that particular operation.

— *Partial response.* A relevant but incomplete description is given. This is particularly dangerous as the Analyst must be able to design a system that covers all foreseeable possibilities.

A successful interview is likely to be one in which the interviewer listens a great proportion of the time, giving the interviewee the opportunity to express himself in an unrestricted manner. As a result the planned discussion area will be comprehensively, but unobtrusively, covered. During the interview the Analyst must also be aware of the body language of the participant. Gestures, facial expressions, degree of eye contact and general posture of the interviewee will convey a great amount of information about values and opinions. These will require responses on the part of the Analyst. For example:

— showing interest and warmth;

— showing encouragement by nodding, smiling or eye contact;

— asking questions if the interviewee appears to falter.

Analysis interviews are concerned with communication. This demands that the Analyst gives the participant both the opportunity and encouragement to express facts, opinions and fears. Communication is a social activity – one-way communication is not true communication (Cherry, 1978).

Concluding the Interview

Keeping to the agreed time is important, and it is usually desirable to arrange a further session rather than to extend the current one. Both participants will have scheduled subsequent tasks and concentration will deteriorate as they become aware of these. The interview should conclude with a brief resume of the main discussion points.

The Analyst will need to write up the discussion points promptly, clearly and, where points of detail or policy are involved, to send a copy to the interviewee for approval. These minutes should help clarify points of misunderstanding or omission as well as identifying responsibilities for further action.

Practice in interviewing is essential. Figure 4.1 shows an interview checklist used by a lecturer in the assessment of role-playing interviews at Leicester Polytechnic. This illustrates the good points of interviewing and could well serve as a checklist for preparing an interview.

Mark:_____

Interviewer:_____   Date:_____

Interviewee:_____   Time:_____

Position:_____   Duration:_____

Planning:	Yes	No	Comments
Was a prior appointment made?			
Was the interview planned?			
Start:			
Was the interviewer punctual?			
Were introductions made?			
Was the purpose of the interview communicated?			
Were roles confirmed?			
Was a request made to make notes?			
Body:			
Were the questions relevant?			
Were questions appropriate to the role of interviewee?			
Were the questions understandable?			
Were answers understood and followed up?			
Were sample documents requested?			
Was any sidetracking controlled?			
Were all procedures and problems covered?			
Conclusion:			
Was interview concluded at an appropriate point?			
Were main points summarised and confirmed?			
Was the possibility of future meeting mentioned?			
Was the departure courteous?			
Conduct:			
Were there any annoying mannerisms?			
Were conclusions jumped to?			
Were responses interrupted by interviewer?			
Were criticisms directed at any member of the organisation?			
Was the plan rigidly followed?			
Did the role of interviewer lapse at any time?			
Was computer jargon used?			

Figure 4.1 Interview Checklist

4.2.3 Questionnaires

When it is impossible because of time, distance or simply cost to interview all the desired people involved in a system, then the Analyst may consider the use of a questionnaire. This is a more structured and formal method of collecting data, but may be the only viable option where there are a large number of dispersed users.

Designing a series of questions which seek out the required information is a skilled activity, and careful planning is required if the results are to prove of any value. There will have to be a balance of open-ended and closed questions. Open-ended questions simply ask a question and leave an adequate space for an unstructured response. The replies may give important insights but they are very difficult to analyse in any systematic or statistical way. An example of an open-ended question is given in Figure 4.2.

What do you consider to be the most significant
problems you face in the operation of the current
booking procedures?

Figure 4.2   An Open-ended Question

In contrast, a closed question is useful for eliciting factual information. This demands that the Analyst has a good understanding of the area under review so that simple, effective questions can be formulated. Each question will have a clear purpose and be posed in an unambiguous way, allowing the respondent to give a definite answer. This may be by means of a selection of answers, actual examples or on a scale of agreement or disagreement. An example of a closed question is given in Figure 4.3.

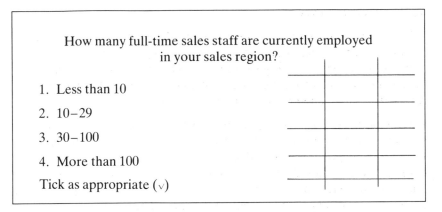

How many full-time sales staff are currently employed
in your sales region?

1. Less than 10

2. 10–29

3. 30–100

4. More than 100

Tick as appropriate ($\sqrt{}$)

Figure 4.3   A Closed Question

Unfortunately questionnaires have several drawbacks which contribute to their low popularity in the investigation of information systems. Low response rates to the questionnaire may lead to unrepresentative responses being analysed to draw incorrect conclusions. The lack of direct contact with the Analyst may mean that questions are interpreted and answered by the respondents in different ways. There is no opportunity to clarify ambiguities unless a follow-up visit or phone call is made. Finally, there is no possibility of the Analyst observing the user's workplace or work practices. The richness of information imparted in the body language of the interview is forsaken.

A few guidelines in questionnaire design are offered below:

— Clearly explain the purpose of the questionnaire.

— Phrase the questions so that they are unambiguous, concise, and unbiased.

— Avoid long questionnaires which seek many different types of information.

— Test the wording and structure of the questionnaire by asking sample respondents to explain their understanding of each question.

— Decide beforehand how the results are to be analysed. A questionnaire can be set out in a way which permits direct input of answers into a computer system, where results can be analysed with an appropriate statistical package.

— Impose a deadline for the response and include a prepaid envelope for postal responses.

It is important that questionnaires are not completely dismissed as a fact-gathering technique. They can be used to 'weed out' a large number of possible respondents so that fact-finding interviews can be better focused. They also have an important role to play in the selection of reference sites, suppliers and the evaluation of software. Significant advantages include:

— It is relatively cheap, particularly when there is a scattered group of users and operators.

— It is free of interviewer distortion and error.

— It permits time to refer to documents and documentation. Questions which concern detailed factual data, eg "How many customers live in the South West?" are suited to questionnaire collection.

— It may be possible to ask more personal and controversial questions particularly if the response is to be anonymous.

A fuller discussion of the relative merits of interviews and question-naires is given in Moser and Kalton (Moser and Kalton, 1971).

4.2.4   Analysing Documents

Documents are very important because they represent the formal information flows of the present system. During all discussions, the documents used by the participants will be referred to, examined and explained. The Analyst should collect specimens of all these relevant documents – input forms, output analyses, reports, invoices, etc, in an attempt to understand how data is passed and used in the present system. Blank forms and copies of completed forms used in the organisation should be collected.

Each document will have its own cycle of creation, amendment, use and deletion. Relevant questions might include:

— what event initially triggers the generation of the document?

— who generates the document?

— how is it prepared?

— where is the data derived from?

— who uses the document?

— for what purpose is it used?

— how is it stored?

— how long is it kept for?

For example, a timesheet recording the hours worked by an employee will be generated every week by that employee, verified by the foreman, used to calculate the wages and to allocate costs to specific projects, and then filed.

The details collected about the document will be guided by the appropriate documentation standards (see later) but features might include:

— when each data item is entered;

— the meaning, size and format of each data item;

— the source of each data item;

— the use of each data item;

— the filing sequence.

Document analysis should also include an assessment of the clarity of the form and how well it satisfies its purpose. The Analyst is particularly looking for ambiguity or obsolescence; for example, column headings that do not correctly indicate the data entered under that item. The volume of documents produced or received is also significant. Any new system must be able to cope with the amount of data passing through it, including seasonal peaks or other variations. A simple average measure is of little use in these circumstances as a system built around averages will underperform for almost half of the time! The growth rate of document use is also important. If the number of documents (say invoices) is increasing by 12% per annum then it would be inappropriate to tailor the system around current volumes.

Documents may also be copied – or be multipart in the first place – with copies going to different people or departments. Each copy must be traced, its purpose ascertained and the filing sequence noted, as this is likely to indicate how old copies are currently retrieved. For example, if copy invoices are filed in date order, then it is very likely that their retrieval will be based on date of invoice, not by any other data item on the invoice such as customer name. This information may be of importance in the subsequent design of files for any proposed computer system. The designer may decide to index on date of order as well as on more likely fields such as customer name or account number.

4.2.5   Special Purpose Record Keeping

Some information which the Analyst requires cannot be obtained by directly interviewing participants or collecting documents used in the present system. This may typically be quantitative aspects such as volumes, trends, frequencies and time intervals. For example, the Analyst may wish to know the relative frequency of different types of enquiry dealt with by a receptionist. It is unlikely that such information would be routinely available, and so a special purpose document would be designed to collect it. Figure 4.4 shows an example of the tally chart kept on a receptionist's desk to record the frequency and type of telephone enquiries from customers.

The most effective special purpose records are those which do not impose an administrative burden on the person required to collect the data. A simple chart requiring a single mark as each event occurs will be completed more accurately than a complex form or instrument which the operator does not fully understand. It must also be recognised that the records refer to a *sample* of events, operations or frequencies, and so every attempt should be made to select a *representative* sample.

Specific opportunities for special record keeping arise when the system is currently computerised. For example, the present pattern of report requests can be logged by writing simple audit programs that collect certain data about the enquiry (such as date, time, who requested it, what was requested, report generation time, etc) but do not interfere with the report enquiry in any way. The pattern of use can then be analysed. Relevant information may also be routinely available from the System Log of the Operating System.

4.2.6   Observation

Observation of a system in normal operation will expose many features which might not be considered relevant, discussed or formally

Customer Telephone Enquiries

Week commencing: 1/6/87

	Request particulars	Arrange viewing	Contracts matters	General enquiry	Others
Monday					
09.00–11.00	卌 I	II		卌 I	I
11.00–13.00	III	I	I	III	II
13.00–15.00	IIII	I			
15.00–17.00	II	I	I	II	I
Tuesday					
09.00–11.00					
11.00–13.00					
13.00–15.00					
15.00–17.00					
Wednesday					
09.00–11.00					
11.00–13.00					
13.00–15.00					
15.00–17.00					
Thursday					
09.00–11.00					
11.00–13.00					
13.00–15.00					
15.00–17.00					
Friday					
09.00–11.00					
11.00–13.00					
13.00–15.00					
15.00–17.00					
Saturday					
09.00–11.00					
11.00–13.00					
13.00–15.00					
15.00–17.00					
Sunday					
11.00–13.00					
13.00–15.00					

Figure 4.4 Special Purpose Record

documented. This may prove useful in gathering information on office conditions which the employees take for granted. A newcomer will notice environmental conditions such as levels of noise, lighting and interruptions. The Analyst may be able to observe the normal levels of supervision and control, the flow of work, the occurrence of bottlenecks in the work flow, the pace of work and the levels of normal and peak workload. Informal systems for producing and storing information, such as personal data files or aids to working, are often discovered only when the Analyst observes a person doing his job. Interoffice communication and handling of spontaneous queries are also more easily observed.

The value of observation as a fact-finding technique depends upon how long the activity is undertaken and the skill of the observer. It is very time consuming and so is best used to supplement other techniques in the building of a picture of the organisation and its information needs. Experience shows that the Analyst may also disturb the work patterns that he is trying to observe and this will naturally reduce the value of the fact-finding exercise.

Finally, it must be acknowledged that observation may be the only feasible method of data collection in certain systems. For example, in a system to control and plan reservoir levels, the pattern of water inflow has to be measured and observed. In such instances, interviewing is clearly inappropriate.

Fact collection requires the cooperation and active involvement of the users of a system. Involvement in this activity can be a strong motivator and promote the acceptance by staff.

4.3 DOCUMENTING THE FACTS

Facts gathered by interviewing, observation, document collection and other techniques will be used by the Analyst in many subsequent tasks. Therefore, they need recording in some standard way for future reference by members of the analysis and design team. Standards for documentation may differ from organisation to organisation, but the examples used below feature the basic elements which need to be recorded.

Standards have a wider use than just documentation. They should aid analysis and design by prompting the Analyst to pose certain questions so that parts of the documentation may be completed. For example,

Clerical Document Specification N C C	Document description POLYTECHNIC APPLICATION FORM		System MIDPOLY	Document 4·1	Name APPLFORM	Sheet 1
	Stationery ref.	Size A4		Number of parts 3	Method of preparation Handwritten / typed	
	Filing sequence Course/Applicant-number		Medium Paper		Prepared/maintained by Applicant - Part A Referee - Part B	
	Frequency of preparation/update On application for entry to course		Retention period 1 year - rejected 4 years - successful 7 years - Maximum		Location School office	

		Minimum	Maximum	Av/Abs	Growth rate/fluctuations		
	VOLUME	100	3000	2000	Peak - August Growing - 15% p.a.		

Users/recipients Course tutor Admin office	Purpose To make decision on suitability for course To record progress of application		Frequency of use

Ref.	Item	Picture	Occurrence	Value range	Source of data
1	Polytechnic - name	X (30)	1		
2	Course - title	X (50)	1		
3	Type - of - course	A	1	S= sandwich F= full-time	
4	Course - start - date	X(10) 9999	1		
5	Main - subjects	X(30)	0-3		
6	Alternative - choice	X(30)	0-1		
7	Applicant - Surname	X(40)	1		
8	Applicant - other names	X(40)	1		
9	Title	A	1		
10	Applicant - previous name	X(40)	0-1		
11	Address - permanent	X(30)	3		
12	Address - correspondence	X(30)	3		
13	Home - tel - no	9(15)	0-1		
14	Corresp. addr - tel - no	9(15)	0-1		
15	Sex	A	1	M= male F= female	
16	Marital status	A	1	M= married S= single	
17	Date - of - birth	99,99,99	1		
18	Country - of - birth	X(30)	1		
19	Country - of - residence	X(30)	1		
20	Nationality	X(20)	1		
21	Date - of - uk - entry	99,99,99	0-1		

Notes
continued on Sheet 2
Ref-no's are marked on specimen document attached

S 41

Author B.W	Issue Date	1 14/5/87

Figure 4.5 Clerical Document Specification

the Clerical Document Specification (see Figure 4.5) requires the Analyst to insert the maximum and minimum number of documents passing through the present system. Thus the Analyst, knowing that such data must be obtained, may include the question in an interview with the relevant user. In this way documentation provides *prompts* for action, not just a *record* of actions.

Standards also represent a method of controlling projects. The head of the systems section can effectively control the way that all system projects are developed by insisting upon completion and adherence to certain standards. This may be very useful because, as Daniels and Yeates once wrote, "until the system is implemented the only tangible evidence that the Analyst has done any work is his documentation". (Daniels and Yeates, 1971.)

Finally, standards impose the use of modelling tools, such as flowcharts, which permit the communication of facts between the Analyst and the users and between the Analyst and his colleagues and managers.

Thus standards are primarily concerned with *aiding* analysis and design, *documenting* the results of analysis and *communicating* these results to other people. The examples described below show documentation to support certain areas of fact finding.

4.3.1   Recording Discussions

A successful interview will trigger a seemingly continuous flow of information, and the Analyst will face the difficulty of recording it all. Much of the information obtained will be summarised and translated into models of the organisation or system. However, many aspects of a discussion cannot be forced into such a format without losing value. Opinions, worries and requirements may all be discussed in an interview, and these need to be minuted to form a permanent record for future reference.

The information will be used in many of the future analysis and design tasks, and the Analyst will need to refer back to past interviews as he begins to develop the models described in the rest of this text. Thus, a data flow diagram (see Chapter 6) will be built up from documented knowledge, while in itself generating the need for further interviews to clarify points, explain issues and discuss details. Discus-

Discussion Record	Title		System	Document		Name		Sheet
NCC	Terms - Reference - Meeting		WOND					

Participants			Date
S. Rogerson (project manager)			14/5/87 2pm
M. Martinek			**Location** Head office

Objective/Agenda		Duration
Agreement on terms of reference amendment		55 mins

Results:	Cross-reference
It was *agreed* that the terms of reference (agreed on 1/2/87) should be amended to include the following item:-	

5. The sales monitoring system should provide the following interrogation facilities.

Specifically:

a. Which salesman is most profitable (in terms of orders and bookings) in a specified month? in a specified year?

b. Which customer accounts fail to meet agreed volume of sales?

c. Which products are successfully promoted by salesmen in the field?

S21

Author
SR

Figure 4.6 Discussion Record Document

sions and the record of those discussions underpin a considerable amount of analysis and design activity. A Discussion Record document (see Figure 4.6) can be used to summarise the meetings that take place as the project progresses.

It is important to stress that the discussion record is not a script! An understanding of the information which has been given is more important that the actual words. A summary in the form of a narrative or a model or picture of some kind is usually adequate, although the level of detail will depend on the purpose of the interview.

4.3.2   Recording Documents

A Clerical Document Specification form is shown as Figure 4.5. Note how it prompts for the information needed to give the Analyst a deeper understanding of the document's use and content. A document description cannot be said to be satisfactorily completed until all the sections of this form are filled in.

This form provides a comprehensive but static view of the data collected on the form. It can be supplemented by a variety of grid charts.

Application.

✻ *Document/Department Grid*

This shows the relationship between the documents and the departments that process them. Document names are entered along the top of the grid and department names down the left-hand side. The numbers (see Figure 4.7) in the intersecting squares show the sequence in which each document is processed by each department. Thus, Delegate List (1) is handled first by the Administrative Assistant who passes it to the Accommodation Officer before it is sent to the Seminar Manager. Nobody seems to use Delegate List (2) at all, so perhaps it is redundant.

Document/Location.

Document/Data Item Grid

This grid is designed to show the duplication of data items on various documents (see Figure 4.8). Document names are entered along the top of the grid and item names down the left hand side.

This form can be used to identify duplication (does the Delegate Telephone Number have to be on the Delegate List?) as well as document omission. Figure 4.8 contains many data items that occur as output (O) but not input (I), thus fact finding cannot be complete. The

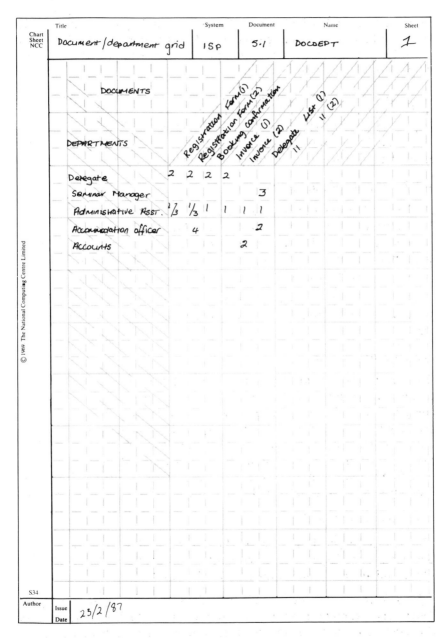

Figure 4.7 Document/Department Grid

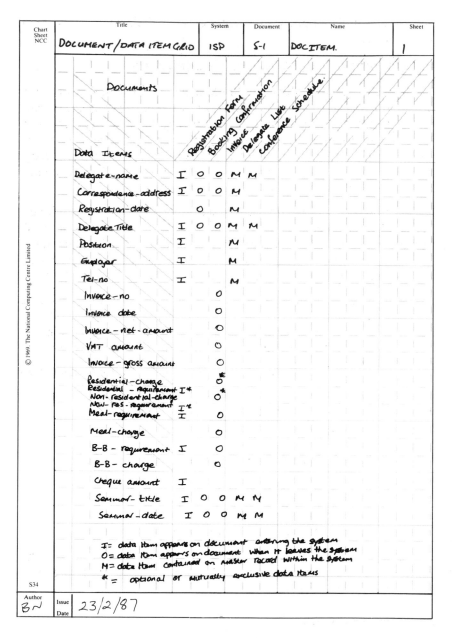

Figure 4.8   Document/Data Item Grid

source of all the input data must be tracked down in subsequent investigation.

4.4 FACT RECORDING: INFOSYS CASE STUDY

InfoSys has a Seminar Manager, Jim McQuith, with overall responsibility for seminars, conferences and short courses. These cover a wide range of information system topics for managers and computer users, including information system design, data security, communications and strategic information planning. The InfoSys Strategic Plan has recognised that this part of the company is very healthy and the company wishes to establish a framework for expansion.

The Seminar Manager currently has only three full-time clerical assistants, Jenny Mathers, Anne Parkes and Gloria Jones. It is their job to undertake all the routine administration involved in the planning of courses and seminars, the registration of delegates and course members, and the preparation of course material.

The Strategic Plan has suggested that the seminar area should be expanded and an appropriate goal has been set (see Chapter 2). It has also been acknowledged that the supporting information systems should be drastically improved, and so a Feasibility Study has been agreed.

The department currently has responsibility for the following activities:

— planning new seminars, conferences and short courses;

— marketing these activities to industry;

— booking delegates on seminars, courses and conferences;

— handling enquiries;

— ensuring supporting material is available;

— sending out billing information to delegates.

A general introduction to the Feasibility Study was given in the previous chapter. Some of the investigations of that study are now examined in greater detail.

Interviewing the Staff

The Seminar Manager, Jim McQuith, was interviewed by the Analyst

Discussion Record	Title		System	Document.	Name	Sheet
N C C	Seminar Booking System					

Participants	Date
Jim McQuith	21ˢᵗ Feb 1987
Dave Nichols	Location
Objective/Agenda	Jim's Office
Seminar booking System objectives	Duration
	2·00 – 3·15

Results: Cross-reference

Findings :

Objectives for the information system were
established as :

   1. To record time, venue and content details of
      forthcoming and past Seminars, conferences,
      and short courses

   2. To record enquiries and firm delegate bookings
      for all seminars, conferences, short courses

   3. To produce marketing information for prospective
      delegates.

   4. To produce information for the administration
      of Seminars, conferences and short courses.

   5. To produce billing information to be sent to
      delegates

Other issues

   1. It was agreed that a meeting would be arranged
with one of Jim McQuith's assistants, Jenny Mathers
Jim will inform me of convenient times

S21

Author

Figure 4.9   Discussion Record Document: INFOSYS

Dave Nichols on 21 February 1987. The main task of the meeting was to identify the objectives of any proposed information system. Figure 4.9 is a discussion record of the interview held in Jim's office.

The interview with Jim McQuith took place in accordance with the recommended 'top-down' strategy. Although Jenny Mathers is responsible for the daily administrative functions associated with the seminar bookings, Dave Nichols had first to establish where the boundary of the discussion should lie. He needed to seek authority and permission to meet Jenny and to gain a perspective of the tasks of booking and administering seminars and other events. The overview obtained from the discussion with McQuith enabled him to plan the interview with Jenny in order to ask more detailed questions.

The initial interview with Jenny Mathers took place in her office. A verbatim transcript is reproduced below.

DN: Good morning, I'm Dave Nichols. Jim McQuith suggested I come to see you this morning.

JM: Oh yes, I was expecting you. Good morning. Well I hope I can help you. Jim said you were going to advise us about getting a computer. All departments seem to be computerised these days.

DN: Yes, perhaps so. Do you mind helping me understand how the booking actually operates?

Telephone rings.

JM: Excuse me a minute, I'll just take this call.

Jenny spends five minutes on the telephone dealing with an enquiry about a forthcoming conference.

JM: Right, where were we?, do you want a cup of tea?

DN: Yes please, white – no sugar.

Jenny returns to the telephone.

JM: Hattie, can you send up two teas please. Both white, no sugar. Thanks . . . Well that's that sorted. Now Mr Nicholas, where were we?

DN: You were going to tell me about the booking procedures.

JM: Yes, it's fairly straightforward really. I'll start at the beginning.

People ring in enquiring about seminars they have seen advertised, or they send in the trade press enquiry slips for details of courses. I like to record all these enquiries as they are people we'd like to put on our mailing list in the future. The problem is that we get so many enquiries that it's impossible.

(some time later)

DN: (after summarising his understanding of the booking system) Well thank you again. You've given me a very comprehensive picture of the documents you handle, the information you would like and the problems you are experiencing. I will go away and let you get back to work. Can we meet again next week to discuss how you send out bills to delegates?

JM: I'm sorry that the phone's disturbed us so much but the office is always like this. Jim is always popping in asking me how well a particular course is booking or how many people attended the last seminar on a particular topic and if anyone couldn't get booked on it. It is difficult to have time to talk without continual interruption . . . Goodbye Mr Nicholas – see you again soon.

The interview with Jenny Mathers was obviously troubled by many disturbances from the telephone and her own manager, and the fact that she had responsibilities which she felt she must get back to. In such circumstances, frequent but short discussions prove more effective than trying to gather all the factual information required in one session.

Analysing the Seminar Registration Form

One of the key documents in the system is the Delegate Registration Form, completed by all delegates wishing to reserve a seminar place. Figure 4.10 is a sample registration form for a recent seminar. The data contents are typical although the style varies according to the seminar or conference topic. Figure 4.11 is the Clerical Document Specification form which Dave Nichols completed after questioning Jenny Mathers about the content and use of the registration form. In this instance, the registration form is a vital input to the system, and hence a thorough analysis is needed to identify data which is collected, but is not used, and data which will be required but is not currently captured.

The registration form consists primarily of personal data completed by the delegate. However, some data items indicate the choice of

⑬ **Automating Systems Development**

⑭ **14-16 APRIL 1987** ⑮

The conference runs from midday Tuesday 14 April to midday Thursday 16th April 1987.

Registration Form

Please return this form no
later than 16th March, 1987 ⑯

It is essential that this registration form be completed by all delegates,
including authors and co-authors.
Please use a separate form or copy for each application.

Please tick boxes
to indicate requirements

Conference Fee (including lunch, refreshments & evening meal)

RESIDENTIAL ☐ £295.00 ①

NON-RESIDENTIAL ☐ £230.00 ②

Accommodation

Evening prior to the course
Monday 13 April: Evening Meal ③ ☐ £ 6.00
Bed & Breakfast ④ ☐ £14.00

*I enclose a cheque made payable to INFOSYS LTD
enclose an order for:

TOTAL £............... ⑤

Signed .... Date .... ⑦

Name (BLOCK CAPITALS PLEASE) .... ⑥

Prof/Dr/Mr/Mrs/Miss* .... ⑧ Position .... ⑨

Employer .... ⑩

Address to which correspondence should be sent: .... ⑪

....

.... Tel No. .... ⑫

*Delete as appropriate

Figure 4.10 Sample Registration Form. Data Items Numbered (see Figure 4.11)

Clerical Document Specification NCC	Document description *Registration Form*		System *ISP*	Document *4.1*	Name *REG FORM*	Sheet *1*
	Stationery ref.	Size *A3*		Number of parts *2*	Method of preparation *Hand*	
	Filing sequence *copy 1 – date of receipt* *copy 2 – delegate – name*		Medium *Glossy paper*		Prepared/maintained by *delegate*	
	Frequency of preparation/update *as required*		Retention period *copy 1 – 6 months* *copy 2 – 2 years*		Location *Admin office*	

	Minimum	Maximum	Av/Abs	Growth rate/fluctuations
VOLUME				

Users/recipients	Purpose	Frequency of use
1. Booking	1, to record booking registration	1, each delegate
2. Accommodation	2, to record accommodation requirements of delegate	

Ref.	Item	Picture	Occurrence	Value range	Source of data
1	Residential fee	✕	1	Y/N	
2	Non-residential	✕	1	Y/N	
3	Evening meal required	✕	1	Y/N	
4	B+B required	✕	1	Y/N	
5	Cheque amount	999v99	1		
6	Delegate – name	✕(30)	1		
7	Date of registration	9(6)	1		
8	Title	✕✕✕✕	1	Prof/Mr/Miss/Mrs	
9	Position	✕(20)	1		
10	Employer	✕(30)	1		
11	Correspondence-address	✕(30)	3		
12	Tel – no	9(10)	1		
13	Conference – title	✕(40)	1	as defined in Programme	
14	Conference date start	99✕(10)9999	2	valid dates	
15	Conference date finish	99✕(10)9999	2	''	
16	deadline date	99✕(10)9999	1	''	

Notes

S 41

Author *D.N*	Issue	
	Date	*23/2/97*

© 1969. The National Computing Centre Limited

Figure 4.1   Clerical Document Specification: Registration Form

accommodation required at the seminar. Reference numbers are marked alongside the data items on the sample registration form.

Collecting Data With a Special Purpose Record

Jenny Mathers was unable to supply Dave Nichols with information on the time she spent on different tasks. She simply told him that she spent all her day doing all her jobs. He decided to design a simple time chart on which she could indicate when she started and finished a task relating to booking. Figure 4.12 shows the tasks chart which she was asked to use for a week. On the left-hand side is a list of all her activities, and the time in quarter-hour intervals is given on the horizontal axis.

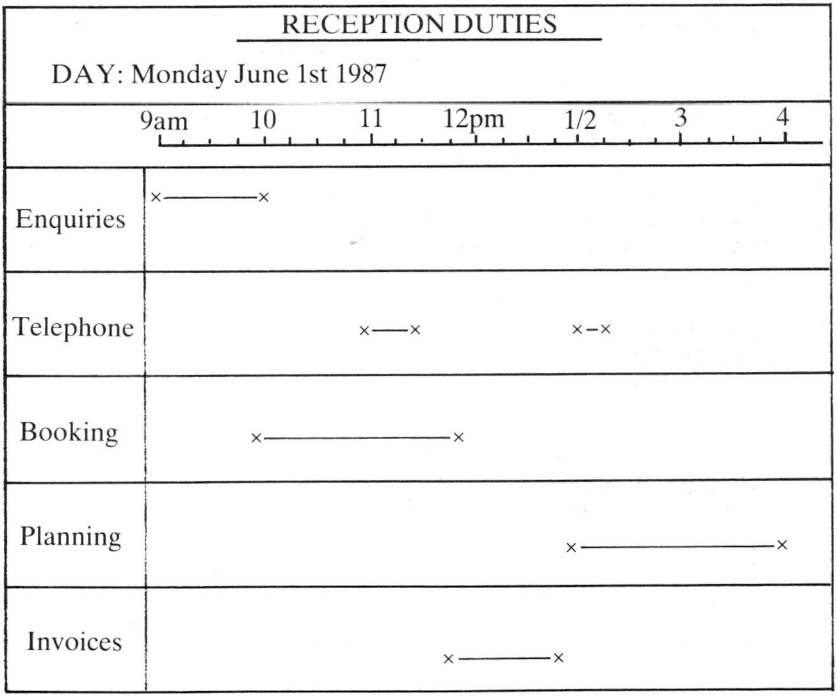

RECEPTION DUTIES							
DAY: Monday June 1st 1987							
	9am	10	11	12pm	1/2	3	4
Enquiries	×——×						
Telephone			×——×		×–×		
Booking		×————————×					
Planning					×—————————×		
Invoices				×———×			

Figure 4.12 Tasks Chart for Jenny Mathers

This record is similar to a planning chart, except that it works in reverse, recording what time she actually spent, not what time she planned to spend.

4.5   OTHER ISSUES

The skills required for fact gathering are benefited by practice and experience and they should be reinforced throughout the learning of the whole analysis and design task. There is a growing tendency, albeit perhaps an unconscious one, to present systems development as the mastery of a set of increasingly complex diagramming models. Kimmerly made this point in a *Datamation* article.

> Current practices in systems analysis have a major deficiency: they place excessive emphasis on the technical details and structured and mechanistic methods, and show a corresponding failure to give due regard to aesthetics, imagery and creativity in the process of systems analysis. (Kimmerly, 1984.)

Many of the technical skills are very useful, but they need to be supported by good interpersonal skills. Communication, negotiation, tact and confidence are all important assets. Confidence grows from practice and experience; hence the need to give the student Analyst the opportunity to undertake interviews and presentations where mistakes and shortcomings can be identified and tackled.

In a survey (Rosensteel, 1981) to determine the importance of various job skills to Systems Analysts, the four highest ranked skills were writing, listening, speaking and interviewing. However, as Parkin *et al* (Parkin, Thornton and Holley, 1987) have pointed out, it pays to look deeply at what is meant by communication skills:

> It would be a mistake to interpret this (communication skill) as verbal fluency or the 'gift of the gab'. For example, one person in our experience had an extreme stutter which made communication in the ordinary sense very difficult. Nevertheless, he was a highly valued Analyst.

They conclude that knowledge of the user area is a crucial foundation for good communication and that an Analyst who knows what he should find is likely to collect important facts even if his communication skills are poor. They suggest that 'good communication' occurs between the Analyst and the user when:

(a) The Analyst has acquired knowledge which the user perceives as valuable;

(b) The Analyst has an understanding of the user's environment and terminology which allows him to ask valuable questions and explain operations and requirements.

These are seen as the strategic requirements of good communication while taught skills such as interviewing techniques are conceived as a technical necessity. Thus the trainee Analyst benefits from studying the business systems that underpin most technical activity.

The successful Analyst must also be seen to undertake tasks in a fair and uncomplicated way. It is likely that he will be privy to information about current organisational performance, eg the 'inefficiency' of the stores department, yet he is expected to gain the confidence of these 'inefficient operatives' to help analyse the problems of the current system. Additionally, he may be expected to design a new system which makes these 'operatives' largely redundant. This presents dilemmas that must be resolved. The Analyst–user relationship is examined in more detail in the companion book *Introducing Systems Design*.

A further difficulty in fact finding is that it is essentially looking for two kinds of fact. The first type exists in the current system and typically is discovered on present documents and in existing procedures. It may be difficult to uncover all the intricacies of the current system but at least they actually exist and can be recognised by the users and operators. In contrast, the second type of fact covers such areas as 'information needs beyond the present system', 'management reporting requirements' and 'management information'. In some instances these may be easy to identify, but in most circumstances it is impossible for users to envisage what might be useful or feasible in advance of receiving it (see Chapter 1). In these areas a Prototype system can be used to experiment and monitor requirements. This might take the form of an application package because it is easier to criticise ("I see no use for that feature") and evaluate ("I did not think that such a report would be possible") a tangible product, than conjure up requirements 'out of thin air'. Groner (Groner *et al*, 1979) gives a case study example where:

> Prototypes were required in the requirements analysis phase because users could not be sure that computer systems were needed, what functions they should perform, or how they would use them.

Attempts to automate the collection of facts date back to the reporting of AUTOSATE, developed by the Rand Corporation (USA), in 1964 (Autosate, 1964), but little reported since. Another early example was Dataflow, where the fact recording was supposed to be done by relatively unskilled clerks using a specification language called Datawrite. This project was abandoned after unpromising field trials. In a recent survey (Parkin *et al*, 1987) it is suggested that fact finding is a worthwhile activity to automate but that few tools actually address this area. Certainly the time spent on this activity, and the recognition of its importance to both Analyst and user seems to demand that automation is given greater attention.

4.6 SUMMARY

This chapter has reflected on the activity of fact gathering. It has:

— introduced a number of specific methods of fact collection;

— described how different types of fact may be documented and recorded;

— illustrated fact collection and recording in the context of the case study;

— raised a number of issues in fact-recording skills that demand further discussion and investigation.

4.7 PRACTICAL EXERCISES AND DISCUSSION POINTS

1   Make a plan of the initial interview with Jim McQuith, the Seminar Manager. Decide on clear objectives for the interview. Conduct this interview with someone playing the role of Jim McQuith and critically analyse the interview afterwards.

2   What fact-finding techniques would you use for investigating the requirements for information to improve the marketing of the seminars and conferences? Which do you think will be most effective? What problems do you anticipate in conducting this investigation?

3   What fact-finding technique or techniques would you use in each of the following circumstances?

— deciding on the design of a new booking form;

— identification of the problem areas in the existing system;

— discovering the information available to the decision takers.

4   Design a special purpose record to determine the volume and type of

enquiries made during a typical day, the complexity of the enquiry and the type of response required.

5 Figure 4.13 is a Conference Booking Schedule which is maintained by the Conference Manager. Document the information contained and make a list of the questions you feel are necessary to complete your understanding of the document and its uses.

6 Reread the interview with Jenny Mathers. Should Dave Nichols have arranged for the interview to be undertaken elsewhere? What practical problems would have arisen if he had and how could these be avoided?

Jenny Mathers twice calls Dave Nichols, Mr Nicholas. He left this uncorrected. Was this a good idea?

7 Discuss the role of application packages in fact finding. What opportunities and dangers do they provide?

8 Investigate and report on the following attempts to automate fact finding:

Autosate (Rand Corporation, USA)
Dataflow (NCC, UK)
Cascade (Trondheim University, Norway)
Ismod (Hein, 1985)

Conference No	Conference/Seminar Title	Date	Target No	Location	Bookings to date
2515	Automating System Development	14/4/87–16/4/87	150	Leicester	~~X~~75
2576	Data Protection Issues	19/6/87	100	Manchester	20
1254	Systems Analysis	18/6/87	20	Manchester	14
1255	Systems Design	21/6/87	20	London	11

Figure 4.13 Conference Booking Schedule

5 System Description Techniques

5.1 INTRODUCTION TO SYSTEM MODELS

A car designer has been commissioned to produce a new luxury sports car. He doodles a few designs on paper and shows them to his colleagues. They make certain comments and criticisms and, as a result, a few changes are made to the design. Finally, the designer is satisfied with three of his designs and so he draws up detailed blue-prints which he gives to the firm's model maker. Scale models of the designs are produced and sent to sales and marketing for customer reaction. The models are also subjected to wind tunnel experiments to investigate the aerodynamics of the design; the results of these tests are used in a computer simulation model to calculate fuel efficiency and drag coefficients.

The designer is using models in at least four different ways:

— The scale model given to sales and marketing is an effective way of communicating his ideas.

— The original doodles represent a clearing of the mind. They are used to generate new ideas and possibilities for the design.

— The wind tunnel experiments show models being used to test ideas and make predictions about the car's performance.

— The computer simulation model uses a representation of the relationship between a variety of variables to compute the car's performance. The model uses previous knowledge and understanding to describe these relationships in a mathematical way. This model assists our understanding of the structure and relationships of the system.

Models are important aids to thinking, creativity and communication and, as these are major preoccupations of the Analyst, their relevance should be clear. The selection of an appropriate model requires flexibility and imagination, adopting a model relevant to the circumstances. The rest of this book is essentially about modelling. It spotlights selected techniques from the dozens available to the Analyst (Martin and McClure, 1984), and further models will be introduced in the companion text, *Introducing Systems Design*. Modelling demands the following conventions and attributes.

Simplicity

Models are used for simplifying the complicated business world. In reality systems are usually large, complex and confusing. If the Analyst is to be able to gain understanding then unnecessary complexity must be stripped away. Recalling all the detail about real systems will also be difficult and the Analyst will wish to keep track of a large amount of information so that it can later be examined and understood. For this purpose pictorial or graphical models are more suitable than narrative ones. Imagine the difficulty of reading a map which is presented entirely in words! A good model represents the system's structures and relationships in a clear and concise way.

Consistency

Consistency is highly desirable. Symbols may be used to represent various aspects of information flow, organisational structure, decision making processes, etc. The same symbol needs to be used consistently through all models. Envisage the difficulties of following a map which changes its scale and symbolic presentation from page to page.

Completeness

Models need to be reasonably complete. Omissions and vague areas of understanding in a model lead to similar difficulties with the actual system. Organisations develop standard models in an attempt to ensure that models used are both consistent and comprehensive. The National Computing Centre Documentation Standards (NCC, 1979, 1987) provide a standard set of consistent and interrelated models which have to be constructed and completed. It includes a cross-reference system which the Analyst can use to check for consistency and completeness in his models.

However, as noted in Chapter 1, it is unlikely that standards alone can provide the variety of models needed in system development. Any attempt to do so can quickly lead to bureaucratic and time-consuming form-filling which threatens the very essence of modelling.

Precision

Appropriate accuracy is necessary if a model is to communicate information effectively. This demands an understanding of the purpose of the model. A program specification is different in detail from an overview of a proposed system presented to user management, and so different models will be necessary. Many user manuals fail to grasp this point. They fail to communicate because all facts are presented at the same level of accuracy and detail.

Hierarchy

Hierarchical models enable the Analyst to maintain several levels of detail in the constructed models. A high-level model will have little detail on it since its primary purpose is to highlight the most important features of the system. A motorway map contained in a diary (Figure 5.1) will show only the motorway network or most of the main roads linking the towns and cities. It will be sufficient to plan a journey from London to Manchester (by motorway!), but it will not help the motorist find his way to a particular district of Manchester. For this purpose a lower level model is required showing the main trunk roads, and this should permit the general location of the desired area. At this point a still lower level model is necessary if the motorist is to find the actual street. He will then consult a street plan, and possibly a location map of the particular premises he wishes to visit to complete his journey.

A similar model is required in information systems design. Each tier of the model will provide vital information, but the different levels will mean that detail can be progressively absorbed. Imagine the confusion of the motorist trying to drive from London to Manchester using only street plans!

This chapter is concerned with models that help the Analyst understand and describe the existing information system procedures. Five specific models are introduced and demonstrated:

— Organisation Charts.

— System Flowcharts.

— Decision Tables.

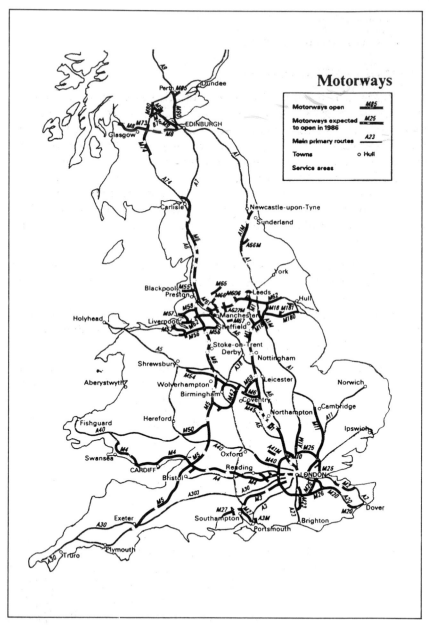

Figure 5.1   A Motorway Map

— Decision Trees.

— Tight or Structured English.

5.2 ORGANISATION CHARTS

Most organisations perform their tasks through an extensive division of labour. The enterprise is divided into many subsections which interact with each other to accomplish some objective or purpose. Information needs to flow between the parts of an organisation to enable the staff concerned to do their jobs. This information system varies from the very formal, highly structured arrangements of large bureaucratic corporations, to the informal flexible practices of many small companies. It is essential for the Analyst to understand the framework of specialised functions and control in the current system and to identify the associated reporting structure.

Organisation charts use simple boxes to represent the division of responsibility and the arrangement of control. In the example shown in Figure 5.2, Departmental Manager A reports to the Chief Executive; four Supervisors report to Departmental Manager A; and a total of 25 staff report to the four supervisors. Similar lines of responsibility can be identified for Department B, C and D.

Departmental splits may be arranged in a number of ways. The chart shown in Figure 5.3 illustrates a conventional division by function. Other possible structures can reflect characteristics such as:

— location, eg division into geographical areas may be used where activities cannot be grouped by other means;

— customer, eg industrial and private customers;

— product, eg division around product lines;

— process, eg division by group of machines.

However, although formal organisation charts may be available in the company, they should not be taken too literally. Indeed, a valuable exercise is to contrast the theoretical arrangement with the one that actually exists. Finding out who actually does what can be very interesting, and it is not uncommon to find quite lowly paid junior staff effectively running departments. This informal organisation structure reflects more closely the arrangement of power and control in an

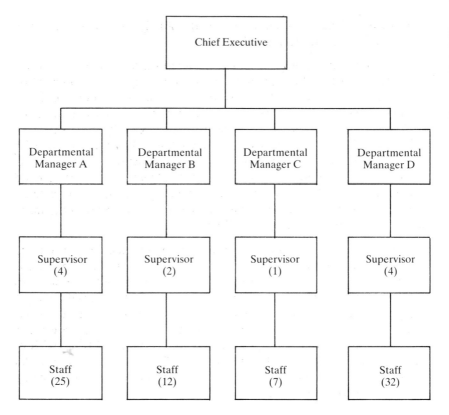

Figure 5.2   Organisation Chart Showing the Vertical Span of Control

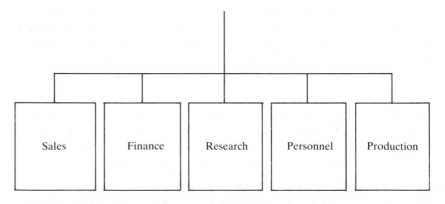

Figure 5.3   Organisation Chart Showing Division of Responsibility by Function

enterprise and hence is important to the Analyst's understanding of responsibility and decision making.

The traditional principles on which organisations have been structured aim to achieve a workable split of authority and responsibility; harmony of objectives (all staff working to a common end); unity of command (each person having only one immediate superior) and unity of direction (one manager and one plan for each objective or set of objectives). Thus the organisation structure has a clear relationship with the Strategic Plan (see Chapter 2). It is difficult to structure an enterprise to facilitate the achievement of the 'common end' if that end remains unclear, blurred or not agreed. Similarly, changes in objectives, or inability to achieve current ones, may lead to corporate reorganisation and realignment of responsibilities.

There has been a wealth of research into organisation structuring which cannot begin to be examined in this text. However, because the structure represents the framework which the enterprise has adopted to achieve its aims, then understanding its formal and informal arrangement is of critical importance to the Analyst. A further practical use has also been identified in the previous chapter. It was noted that interviews should be undertaken in a top-down manner, and indeed Dave Nichols interviewed Jim McQuith before his subordinate Jenny Mathers. An appropriate organisation chart is an obvious prerequisite to the planning of such a series of interviews.

5.3 SYSTEM FLOWCHARTS

It was noted in the previous chapter that the analysis of documents is an important task because the existing forms and reports represent the information flow and contents of the current system. A detailed form was introduced (Clerical Document Specification (see Figure 4.5)) to record the structure of documents used in the present system.

However, this standard form gives a relatively static view of the data, failing to clearly show the sequence and interaction of documents. Understanding these latter dynamic aspects of the system is important because any replacement system is likely to affect and change such flows and so the Analyst must first be sure that they are properly understood.

A commonly used tool for documenting current flows is the system flowchart. This summarises what operations are undertaken and where and when they take place.

The system flowchart is divided into columns with the name of each department, function or personnel written at the top. Inputs from outside the context of the flowchart are shown to the left and outputs to the right. Symbols representing the operations undertaken and the documents used are then placed in the appropriate column. Decisions which lead to different actions being taken can also be shown.

NCC system flowchart symbols are widely used and recognised and these are shown in Figure 5.4.

The first step in constructing this model is to identify the main departments or functions involved. Appropriate symbols are then drawn in a sequence which gives a general flow of data from top to bottom and left to right. Arrows are used on the connecting lines to indicate the logical flow or sequence where the flow is not in the standard direction. No interaction is implied by crossing lines. The model can then be completed by cross-referencing some or all of the symbols to other charts or documents (such as the Clerical Document Specification). Cross-references can be given in the top stripe of the symbol, whilst the physical characteristics of the input, output or storage may be shown by labelling the bottom stripe.

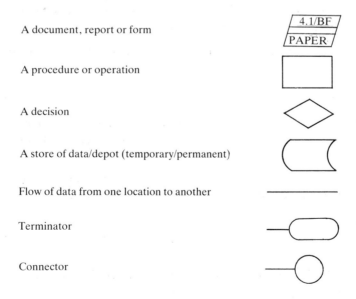

A document, report or form

A procedure or operation

A decision

A store of data/depot (temporary/permanent)

Flow of data from one location to another

Terminator

Connector

Figure 5.4   Flowchart Symbols

Figure 5.5 is a system flowchart for admissions procedures in Computer Science at a college. The responsibilities of such people as the course tutor, the admissions tutor and the general office are clearly shown by the position of the symbols. The flow of documents, such as the application form, through the system and the storage of data at different points is also illustrated.

These models are relatively easy to follow and show the sequence of operations in a clear way. They are also fairly intelligible to users and can be employed to support the Analyst in his explanation of the current operational systems. The system flowchart also highlights such issues as excessive information flow, duplication of information stored in different locations, and reasons for delays in particular operations.

A number of guidelines can be offered for drawing system flowcharts:

— Decide on the column headings before drawing the flowchart. Do not add them as you go along.

— Try to show flows moving from top to bottom and from left to right.

— Attempt to keep flow continuous and self-explanatory. This may demand the addition of explanatory processes which help to illustrate the timing of subsequent actions.

— The system flowchart is a summary model. Connectors can be used to link pages but it is desirable to keep the model relatively small. It is supposed to simplify and clarify – not just to give a visual version of a long narrative.

— There is no 'one best way' of drawing the model. A number of flowcharts, correct in logic but differing in detail, are likely to be equally valid.

— A certain amount of confusion is caused by an overlap between a decision and a process. Which of the following is correct?

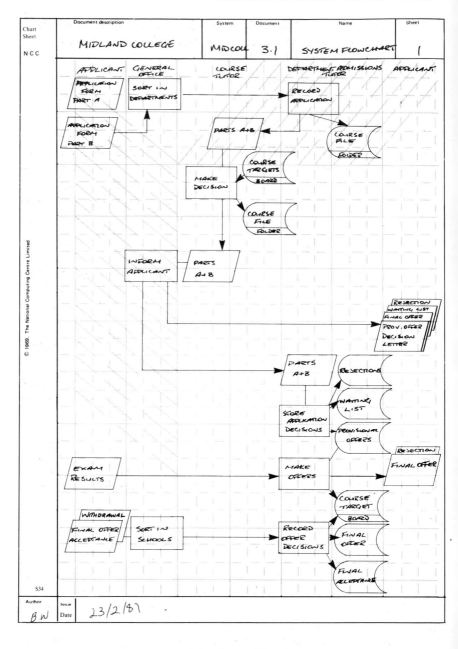

Figure 5.5   System Flowchart

The answer is that both are acceptable, but that each will be appropriate in different circumstances. The logic of the chart is more important than the shape of the boxes.

— All 'loose ends' should be tidied up. So if there are three parts of an order form each should end up at some clearly defined point – probably in an appropriate permanent file.

Some Analysts have discarded this model in recent years in favour of other techniques such as data flow diagrams – discussed in detail in the next chapter. However, whilst acknowledging some problems associated with its construction, it is felt that the clarity and simplicity of the system flowchart make it a most useful tool for describing the flow of operations and control in the existing system.

5.4 DECISION TABLES

It is inappropriate to record all the detail of the current system on a system flowchart. More complex circumstances, particularly those where several criteria determine an action, demand more specialised models.

The following narrative summarises an interview with an accounts clerk in a Mail Order Book Club.

Priority treatment is given to members who order more than £100 value of books in a year and have a good payment record, or who have been members for over 10 years.

The statement is ambiguous as it is not clear if poor paying members of over 10 years' standing will receive priority treatment. Drawing a decision table will highlight this ambiguity and the Analyst will be able to express the logic precisely. The following narrative is now more explicit but is hardly conversational!

Members of the Book Club whose order value exceeds £100 in a year and have a good payment record are given priority treatment. Where a membership has been over 10 years the member will get priority treatment if the order value exceeds £100 per annum, in which case the club is prepared to overlook a poor payment record. If the order placed by a member of more than 10 years' standing does not exceed £100 during the year, but the member has a good payment record, then priority treatment is also given. In all other circumstances the order is given normal treatment.

A decision table is useful for representing the conditional logic of processes where different actions are taken depending upon the occurrence of a particular combination of circumstances. This represents typical discussions where staff explain that "when x and y happen then we will do this and that, on the other hand y might not happen and, in this case, we will do something else . . .". In such situations the Analyst will often feel confused and unsure that the rules of the enterprise have been completely understood. Expressing the logic in the form of a decision table should clarify the rules governing the options.

5.4.1   Drawing a Decision Table

A decision table is divided into the four parts, as shown in Figure 5.6. The conditions, that determine which actions will result, are listed in the condition stub. Combinations of these conditions are then identified and expressed as rules or condition entries. The possible actions which can occur as a result of different condition combinations are listed in the action stub. The decision table is completed by entering the relevant actions for each condition rule identified.

The simplest type of decision table is a limited entry decision table in which conditions are expressed as questions that may be answered by a simple Yes (Y) or No (N). The condition entries or rules are then specified as combinations of these answers. This type of table is self-checking to the extent that there is one rule for each possible combination of conditions.

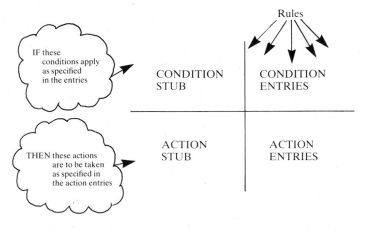

Figure 5.6   Structure of a Decision Table

The relevant action for each combination of conditions is recorded by an X in the action entries section.

The construction of a limited entry decision table can be illustrated from the example Book Club narrative.

1 Identify all conditions, being careful not to include mutually exclusive conditions such as Good Payment Record and Bad Payment Record. Write the conditions down in the Condition Stub with the most significant or critical one first. If more than four conditions apply then divide the table in two.

c1: Good payment record?

c2: Order value > £100 pa?

c3: Member > 10 years?

2 Identify all actions possible and record them in the Action Stub in the sequence in which they occur.

a1: priority treatment

a2: normal treatment

3 In the case of a limited entry table, the number of rules can be calculated by using the formula (2^c). Ensure that all the condition entries or rules have been expressed. In this example, three conditions generates 2^3 rules, since each condition entry can be Y or N. This explains why the division of a table at four conditions was suggested. 2^5 gives 32 rules, which may be a little unwieldy. A suggested format for allocating the Ys and Ns is given below. For Z rules, the first row has Z/2 Ys and Z/2 Ns, reducing down until the last row which is always YNYNYN etc.

c1: Good payment record?	Y Y Y Y N N N N
c2: Order value > £100 pa?	Y Y N N Y Y N N
c3: Member > 10 years?	Y N Y N Y N Y N
a1: priority treatment a2: normal treatment	

4 Action Entries are now made. These are derived by applying each

combination of conditions to the actions described in the narrative.
The following table is produced.

| c2: Good payment record? | Y | Y | Y | Y | N | N | N | N |
| c1: Order value > £100 pa? | Y | Y | N | N | Y | Y | N | N |
c3: Member > 10 years?	Y	N	Y	N	Y	N	Y	N
a1: priority treatment	X	X	X		X			
a2: normal treatment				X		X	X	X

5   In limited entry tables there should be 2^c rules in total, where c
is thc number of conditions expressed in the Condition Stub. Check
the table completeness by counting the rules.

6   Check the table for redundancy. This is present if two or more rules
exist with different combinations of conditions leading to the same
actions. In such circumstances consolidate the rules where possible.
A consolidation is shown in the Condition Entries by a broken line.
This effectively indicates that this condition is irrelevant – it can be
Yes or No – the action remains the same. Two consolidations have
been undertaken to produce the following table.

| c2: Good payment record? | Y | Y | Y | N | N | N |
| c1: Order value > £100 pa? | Y | N | N | Y | Y | N |
c3: Member > 10 years?	–	Y	N	Y	N	–
a1: priority treatment	X	X		X		
a2: normal treatment			X		X	X

7   It is possible to place an ELSE statement at the end of the chart to
permit simplification and conciseness.

c1: Good payment record?	Y	Y	N	E
c2: Order value > £100 pa?	Y	N	Y	L
c3: Member > 10 years?	–	Y	Y	S
				E
a1: priority treatment	X	X	X	
a2: normal treatment				X

8 Use a top-down hierarchy of tables where more than four conditions are involved. This is useful in circumstances where there are too many conditions operating to result in a simple table. The most important conditions are evaluated in a high-level table and, depending upon the responses to these, the flow is directed to one of several other decision tables. An illustrative example is given in Figure 5.7.

Table 1

Places available on course?	Y	Y	N
Financial support available?	Y	N	–
Go to Table 2	X		
Go to Table 3		X	
Reject application			X

Table 2

Qualifications satisfactory?	Y	Y	N	N
Reference satisfactory?	Y	N	Y	N
Make offer	X			
Reject application				X
Go to Table 3		X	X	

Table 3

Interview satisfactory?	N	Y	Y
Mature student?	–	Y	N
Offer supported place		X	
Reject application	X		
Place on waiting list			X

Figure 5.7   Hierarchy of Decision Tables for Student Application

5.4.2   Extended Entry Decision Tables

The concise nature of the decision table has prompted its extension to include circumstances where the Condition Entries are expressed as values of some kind. An extended entry decision table may also include symbols or codes in the action entry section of the table.

Consider the following limited entry decision table (Figure 5.8).

Member < 1 year	Y	N	N	N
Member 1–5 years	–	Y	N	N
Member 6–10 years	–	–	Y	N
No discount	X			
Discount 10%		X		
Discount 25%			X	
Discount 50%				X

Figure 5.8   Limited Entry Decision Table

Using a limited entry decision table for this example appears unsatisfactory since the conditions are all related to each other. Only one action can take place and the table looks unnecessarily complex. Figure 5.9 describes the same logical procedure in a simpler way by using an extended entry decision table.

Length of membership	< 1	1–5	6–10	>10 years
Discount %	0	10	25	50

Figure 5.9   An Extended Entry Decision Table

Hybrid decision tables can also be constructed where the conditions are a mixture of Yes and No and other values as in Figure 5.10.

Good payment record	Y	N	Y	N	etc ...
Length of membership	< 1	< 1	1 < 5	1 < 5	
Accept	X		X		
Reject		X		X	etc ...
Discount %	0		10		

Figure 5.10   A Hybrid Decision Table

5.5   DECISION TREES

Decision trees are another way of showing the alternative actions that can result from different combinations of circumstances. The diagram resembles a fallen tree, with a root on the left-hand side and branches representing each decision. The tree is read from left to right and the actions to be undertaken are recorded down the right-hand side of the diagram. The conditional logic of the order handling process described in the previous section has been recorded in the form of a decision tree in Figure 5.11.

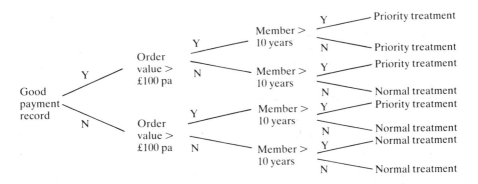

Figure 5.11   The Order Handling Decision Tree

5.6   TIGHT OR STRUCTURED ENGLISH

Ordinary English narratives often lead to misunderstandings due to the ambiguity of the language. The narrative can be sharpened by using tight or structured English. This uses standard narrative constructs that constrain the language but, at the same time, permit a certain degree of user understanding. The structure is enhanced by indentation.

Book Orders

 Get Member Order details

  IF Good Payment Record

   IF Order Value > £100 pa

    DO Priority Order

   ELSE (Order Value < £100 pa)

   IF Member > 10 years

    DO Priority Order

   ELSE (Member < 10 years)

    DO Normal Order

   ENDIF

  ENDIF

  ELSE (Poor Payment Record)

   IF Order Value > £100 pa

   IF Member > 10 years

    DO Priority Order

   ELSE (Member < 10 years)

    DO Normal Order

   ENDIF

   ELSE (Order Value < £100 pa)

    DO Normal Order

   ENDIF

  ENDIF

Processes may be described in structured English by several basic constructs:

Sequences of events or actions are described by simply using the top to bottom order of the statements. For example:

Accept application.

Store applicant data.

Add to list of applicants.

Send letter of acknowledgement.

The four actions above are undertaken in the order presented in this sequential structure. The statements are presented in line and all actions are performed.

Circumstances where actions depend upon differing conditions can be described using conditional statements such as IF ... Else ...ENDIF.

Where one of several possible cases, apply statements such as:

CASE 1 ...

CASE 2 ... etc
or
WHEN ... DO

WHEN ... DO

Repetition of action can be described using statements such as:

DO WHILE ... ENDWHILE

or DO UNTIL ... ENDUNTIL

or FOR ALL ...

For example:

DO WHILE still forms on pile

Read application details

IF application is for overseas student

Request proof of financial support

ELSE (home student)

ENDIF

ENDWHILE

```
    IF course has vacancies

        CASE (exam results meet requirements)

            Make final offer to applicant

        CASE (results not known)

            Make a provisional offer

        CASE (results not good enough)

            Reject application

        ENDCASE

    ELSE (course full)

    ENDIF
```

This clearly shows the hierarchy of constructs (shown in capitals) and comments have been added to clarify the narrative. Structured English will be encountered again in the later chapters of this book and in the companion text *Introducing Systems Design*.

5.7   FACT DESCRIPTION TECHNIQUES: INFOSYS

At a subsequent meeting with Dave Nichols, Jenny Mathers describes the normal procedure adopted on receiving a Booking Form:

> When we get forms in through the post I check that all the items have been completed. It's surprising how many people send in a form without details of their organisation, so we can't bill them, or they may try to book seminars which are not adequately described so it is not clear which one they want booking on. I will return any booking not completed correctly with a letter asking for clarification or further details.

> If there is space on the seminar which they wish to attend then the booking is confirmed to them in a standard letter and their name and other details added to the delegate list. If the seminar lasts more than a day then the Accommodation Section needs to know and so a request is completed and sent to them. When they have booked the room and meals I receive a confirmation of this. Occasionally we cannot cater for all those wishing to attend on a certain date. In this case a letter is sent offering alternative dates. The delegate quite rightly gets upset if his dates are changed once

these have been confirmed, but this is necessary; we can't have the situation where they are scattered around several hotels in the district.

If there is no space on the seminar requested then we will add their names to a reservation list, as some people do drop out later due to business or personal reasons. If the booking is for a block booking, of over 12 people from one organisation, then I consult Jim (the Seminar Manager) who plans and schedules the seminars as special arrangements may be made for such customers.

Each delegate who attends a seminar is given an attendance certificate for claiming his own expenses. The accounts department sends an invoice directly to the sponsoring organisation and keeps a copy filed for matching with the payment later. Where the delegate has prepaid the booking the invoice is raised for accounting purposes only and is not forwarded to the delegate unless requested.

5.7.1 Using the System Description Techniques

The information concerning the present operation of the seminar booking system can be recorded in both summary and detail. For example, the division of responsibilities and the control structures can be described on an organisation chart (see Figure 5.12a). InfoSys has relatively few staff involved in the seminar section, so it is easy to represent the structure in the organisation chart. Although the present investigation is concerned only with the seminar booking system, drawing this particular chart will ensure that the place of the seminar section in the larger organisation is appreciated. It is not necessary to complete all sections of the chart in great detail, only the area involved in the system which is being investigated needs this. Figure 5.12b gives the proposed organisation structure after the implementation of the recommendations of the Strategic Plan.

A system flowchart describing the functions of the departments involved in the seminar booking system is given as Figure 5.13. With such an overview model it is difficult to strike a happy medium between what should be left in and the detail that must be discarded. Successful drawing of these models will require further discussions with staff to obtain extra information, confirmation of procedures and further administrative details.

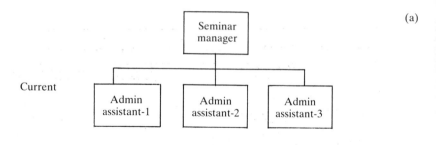

Current

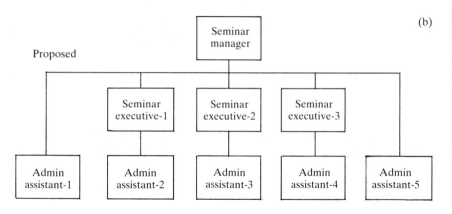

Proposed

Figure 5.12   The Infosys Organisation Chart

Drawing the system flowchart of the seminar booking system raised a number of problems of clarifying procedure and deciding upon appropriate representation.

— Are telephone bookings ever accepted (where a booking form is not completed)?

— Several important decisions determine the flow within the system:

• Is the seminar full?

• Is it a block booking?

• Is accommodation required?

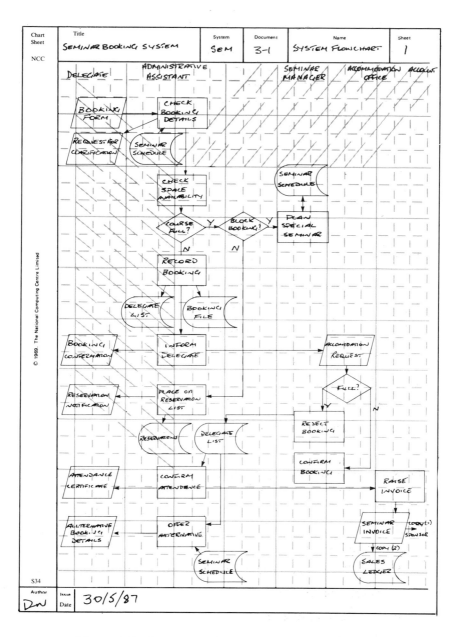

Figure 5.13   System Flowchart: Seminar Booking System

These can be represented alternatively using process boxes in preference to decision symbols. It can be argued that the detailed conditional logic involved requires the use of models such as decision tables or decision trees which are more suited to this purpose.

— Some files need to be repeated for clarification reasons but the need for showing the same files used by processes under more than one heading indicates the possibility of conflicting access requirements.

— The process 'confirm booking' does not have any resulting document flow and hence the Analyst needs to clarify his knowledge in this area.

The processing logic involved in the acceptance of bookings from delegates is relatively complex. This needs to be described in more detail than is feasible on the system flowchart. A decision table or decision tree may be used, dependent upon personal preference. These two representations are developed and shown in Figures 5.14 to 5.17.

Booking form completed correctly?	Y	Y	Y	Y	N	N	N	N
Seminar space available?	Y	Y	N	N	Y	Y	N	N
Block booking?	Y	N	Y	N	Y	N	Y	N
Return booking form with query						X	X	X	X
Confirm booking	X	X						
Place on reservation list				X				
Enquire with seminar manager			X			X		

Figure 5.14   The Booking Acceptance Decision Table

Consolidation yields the decision tables shown in Figures 5.15 and 5.16.

Booking form completed correctly?	Y	Y	Y	N	N	N
Seminar space available?	Y	N	N	Y	N	N
Block booking?	–	Y	N	–	Y	N
Return booking form with query				X	X	X
Confirm booking	X					
Place on reservation list				X		
Enquire with seminar manager		X			X	

Figure 5.15   A Consolidated Seminar Booking Decision Table

Booking form completed correctly?	Y	Y	Y	N	E
Seminar space available?	Y	N	N	N	L
Block booking?	–	Y	N	Y	S
					E
Return booking form with query				X	X
Confirm booking	X				
Place on reservation list			X		
Enquire with seminar manager		X		X	

Figure 5.16   A Consolidated Seminar Booking Decision Table Using the Else Rule

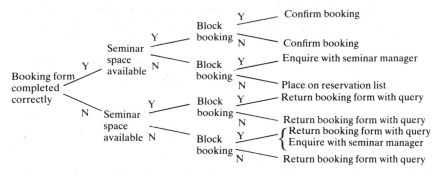

Figure 5.17   The Booking Acceptance Decision Tree

This decision tree is rather lengthy and can be consolidated to the simpler decision tree shown in Figure 5.18.

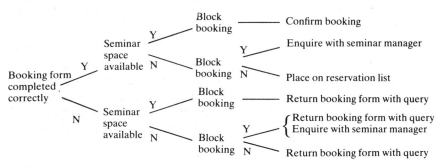

Figure 5.18   The Booking Acceptance Decision Tree Showing Consolidation

5.8 SUMMARY

Decision tables, decision trees and structured English are essentially tools for logic or process description. In a comparison that also distinguishes structured and tight English – a distinction we have chosen not to pursue – Gane and Sarson (Gane and Sarson, 1980) assess these tools against a set of criteria. They conclude that decision tables are very good from the perspective of logic verification and computability but that they are not easy to use or amend. In contrast, decision trees are simple, good at displaying the logical structure, but poor from the viewpoint of verification and computability. They arrive at the following conclusions:

Decision trees are best used for logic verification or moderately complex decisions which result in up to 10–15 actions. They are also useful for presenting the logic of a decision table to users.

Decision tables are best used for problems involving complex combinations of up to five or six decisions. Large numbers of combinations of conditions can make decision tables unwieldy.

Structured English is best used wherever the problem involves combining sequences of actions with decisions and loops. (Gane and Sarson, 1980.)

The computability of decision tables makes them a very attractive candidate for automation. One commentator (Lew, 1984) has claimed that program code derived automatically from decision tables can be proven correct and hence do not need recasting into conventional languages or flowcharts. A much earlier advocate of the decision table (Fergus, 1969) gave examples of four preprocessors, including a 1967 product producing PL/1 source statements. Decision tables also had a central role in Grindley's Systematics specification language (Grindley, 1968).

System flowcharts are best viewed as complementary to the logic definition techniques. They can show decisions but not very convincingly. They are currently rather unfashionable and this has probably contributed to an absence of automated versions. Parkin *et al* (Parkin, Thornton and Holley, 1987) conclude that it is both beneficial and feasible to automate system flowcharts and have embarked upon a prototype.

This chapter has concentrated upon the existing physical system. The models have been introduced in this context although some, notably the logic definition tools, will also be used elsewhere. Understanding current arrangements is important but must be tempered with caution – we must not drown in the "current physical tarpit". Consequently, it is now necessary to move to a different level of modelling, that of logical system definition. In some respects this will be the first positive step in the design of a replacement system.

Thus, in summary, this chapter has

— Introduced the desirable features of system models.

— Described the purpose, construction and relevance of a series of models that can be used to describe the current organisational systems. Further case study examples of some of these models were constructed.

— Briefly contrasted the strengths and weaknesses of each model.

5.9   PRACTICAL EXERCISES AND DISCUSSION POINTS

1   Fabrics Galore Ltd has a standard method of receiving yarn stock requests. As requests are phoned through from the factories, a sales clerk checks to see if any stock is held. If no stock is available then a phone call is made to the supplier requesting a special delivery and the stock request is filed in the 'HOLD' file. If stock is available, then the clerk checks to see if the stock meets the requirements of the works order. If stocks do not meet the works order requirements, the clerk checks to see if there is a purchase order outstanding on the yarn stock. If such a purchase order has already been made then the clerk sends a postcard requesting urgent delivery. In both cases, the available stock is sent, although it comprises only part of the order, and the balance is adjusted accordingly. The stock request is then filed in the 'HOLD' file so that the rest of the order can be fulfilled when yarn stocks are replenished.

However, if the yarn stocks meet order requirements, the sales clerk examines the stock to see if they are below reorder level. If they are below this level then the clerk must check to see if a purchase order is outstanding. If there is an order outstanding then a postcard is sent demanding urgent delivery, but if no order has been made then a purchase order form is made out and despatched to the yarn supplier. If the balance is not below the reorder level then these actions are not required. In all cases when the stock is sufficient to meet orders the whole batch of the order is despatched and the stock balance is adjusted accordingly.

(a)   Draw a decision table to record the above procedures.

(b)   Suggest alternative methods of representing the above rules.

2   Discuss the relative merits of system flowcharts, decision tables and decision trees. Which do you find easiest to draw and understand?

3   City Dyers Ltd is a company which specialises in the dyeing and finishing of fabrics for the local knitting industry in Leicester. The management are investigating the feasibility of a computer system to monitor the progress, costs and quality of producing customer orders in the dye works. The company has expanded rapidly since it was established in 1982 and two sister companies of a similar size have been recently acquired to cope with a healthy customer order book.

CITY DYERS LTD

TK *17903*

FROM/KNITTER

CD

ACCOUNT_____
DISPATCH TO

E.G JONES
HINKLEY

Shade *Navy* ____ No. of Rolls *2* ____

Quality *Softened Poly Rib* ____

Special instructions *Part lot* ____

Date *24/1/87* ____

Delivery Note No.____

Invoice No.____

Required Width *26"* ____

PIECE No.	WEIGHT IN		WEIGHT OUT		METRES	YIELD	REMARKS
	Kgs	Gms	Kgs	Gms			
154	*15*	*2*	*14*	*9*			*26" finish*
382	*15*	*7*	*15*	*3*			
TOTAL							

Date dyed *24/1/87* __ Date finished *24/1/87* __ Finishing Operator *SM* ____

Signature *AB.* ____

Figure 5.19 A TK Note

CITY DYERS LTD

FROM:

E. G JONES

DELIVER TO:

E.G JONES
HINKLEY

ORDER No.: *1394*   DATE: *21·1·87*

PIECE No.	GROSS WT. IN		FINISHED WT	
	Kg	Gms	Kg	Gms
154	15	2		
395	12	8		
382	15	7		
391	15	2		
403	15	3		
139	15	3		
145	13	1		
	102	6kgs total		

Lot No.   *19*
Shade   Navy / sky
Width to Finish   *26"*
No of Rolls   *7*
Type of Fabric   Poly Rib
Quality

SPECIAL INSTRUCTION

Finish 26"
approx 7 rolls

30 kg navy
rest sky

SIGNATURE: *S. Williams.*

Figure 5.20   A Customer Order

Up to now very little paperwork has been necessary but the directors realise that growth produces its own problems and the continued custom of some companies will depend on high quality assurance and speedy service. The sharing of orders between the three companies is envisaged if a quick turnround is required and appropriate resources are available at one of the other sites. However, in such circumstances the customer should always be invoiced by the company receiving the order in the first place and intercompany billing will redress the balance.

The present procedure for processing orders is based on a 'TK note' (Figure 5.19) which is derived from details on the customer order (Figure 5.20) and issued for each batch of rolls of a particular fabric-type (eg cotton, polyester) to be dyed one particular shade. There are approximately 30 fabric types but the number of shades is virtually infinite. About 25 customer orders are currently received every day each consisting of 10–20 rolls/pieces weighing an average of 11 kilograms each. In some cases the customer will instruct City Dyers to dye given proportions of an order (indicated by weight) in different shades to their specification. In these circumstances the company generates several TK notes for the order. Brief details (eg customer name, date, advice number, number of rolls, weight, yield required, shade and fabric quality) are recorded in a daybook (Figure 5.21) which acts as a log of work undertaken by the company.

The dyehouse manager agrees and records on a customer shade card (see Figure 5.22) the shades of the fabric on order. This is agreed with the client prior to receipt of the order on the basis of a sample of fabric, specifically test-dyed for this purpose. The recipe for a particular fabric shade specifies the quantities of several dye types and the time they are to be applied. There are also likely to be special finishing instructions (such as the addition of softener) or a specification of a 'required width finish'. The processes are relatively simple and consist of washing at various temperatures, several stages of dyeing, softening, stretching or compressing under a heat treatment and final finishing to the customer's instructions. Only the machine to be used for the dyeing process is recorded on the shade card as the other processes may only be undertaken on one machine.

The shade cards for TK notes currently being dyed and finished are displayed, in the order they are to be processed, on a board in

CITY DYERS LTD   DATE: JANUARY 1987

CUSTOMER ORDER NO.	CUSTOMER	DATE	TK No.	WEIGHT (kg)	WIDTH	SHADE	FABRIC TYPE	INVOICE No.	DATE	REMITTANCE
1394	E.G. JONES	21	1903	7 30 0	26"	NAVY	POLY RIB	639	24/1/87	
1594	"	21	1904	7 72 6	26"	SKY	POLY RIB	639	24/1/87	
779	MUSANI GARMENTS	21	1905	32 245 4	14"-16"	WHITE	NYLON	640	24/1/87	
779	"	21	1905	22 246 6	14"-16"	SKY	NYLON	640	24/1/87	
733	SMART KNIT	22	1907	2 19 3	24"	SKY	NYLON	641	12/2/87	
1302	SUNNY TEXTILES	22	1908	16 344 2	68"	BEIGE	ACRYLIC/COTTON	622	6/2/87	
1286	BEACHWEAR MFG	22	1909	13 127 5	28"	WHITE	NYLON			
1286	"	22	1910	15 144 5	28"	PINK	NYLON			
1286	"	22	1911	13 143 4	28"	SKY	NYLON			
1716	ADAM + CO	22	1912	9 1943 3	66"	OPTIC WHITE	SPUN POLYESTER	619	26/1/87	
1715	"	22	1913	12 2242 4	66"	OPTIC WHITE	SPUN POLYESTER	619	26/1/87	
1714	"	22	1914	43 2704 4	66"	OPTIC WHITE	SPUN POLYESTER	619	26/1/87	
1373	SMART KNIT	22	1915	12 125 1	22"	SKY	NYLON			
1606	STARWARD FABRICS	23	1916	15 271 65	206cm	PINK	POLY RIB	607	26/1/87	
1606	"	23	1917	18 272 8	206cm	FOREST GREEN	POLY RIB	607	26/1/87	
1606	"	23	1918	24 351 9	206cm	BLUE	POLY RIB	607	26/1/87	
1392	CRAVEHURST/KIMBELLE	23	1919	13 230	60"	SKY	PEEL BACK COTTON	642	27/1/87	
1794	E.G. JONES	23	1920	8 106 2	30"	LEMON	TERRY TOWELLING	639	24/1/87	
1795	"	23	1921	8 103 1	30"	OPTIC WHITE	TERRY TOWELLING	639	24/1/87	
1244	LAKHI JERSEY	23	1922	4 61	61"	WHITE	SINGLE JERSEY	617	3/2/87	
1717	ADAM + CO	23	1922	21 423 0	64"	PADIWHITE	SPUN POLYESTER	619	28/1/87	

Figure 5.21   The Sales Daybook

CITY DYERS LTD			Process		
Customer *E.G JONES*					
TK no *17903*	Machine		*26" finish*		
Shade *NAVY*	Type *POLY RIB*				
Weight *30 Kg*	Rolls *2*		*Softened*		
Dyes/Chemicals	Qty/Time added				
Art Blue	*40g*	*75Mm*			
Dysol 40	*25g*	*2Mm*			

Figure 5.22   The Customer Shade Card

the works office. Employees (organised into two eight-hour shifts), must refer to the display board for instructions on the next stage of processing. This will be found on the TK note which accompanies the fabric through the dyeworks.

Minor clerical errors have led the company into difficulties in keeping consistent recipes for customers who send repeat orders. This has resulted in having to redye customer fabrics at the company's expense and, in a few cases, scrapping the whole fabric completely. There are also occasions when the dye stock itself is insufficient to meet customers orders in time. Unfortunately, there is very little warehouse space at each site and customers' orders are required to be delivered or collected as soon as possible after finishing. Dealing with customer enquiries regarding the status of completion of a particular order has, to date, relied on the memory, knowledge and experience of employees.

When the whole of the fabric on an order is finished the customer is invoiced on the basis of weight at a previously contracted price per kilogram (gross fabric weight). Each bill may refer to a maximum of 30 different orders although the average is five.

Specimen documents currently used in one of the factories are shown in Figure 5.19 to 5.22.

The company wishes to obtain better information on:

(a)   the current location and work status (received, finished, started, delivered) of a particular order;

(b)   the recipes for particular customer fabric shades;

(c)   the outstanding customer orders to be billed;

(d)   the dye cost of particular lots undertaken at each factory (it is envisaged that labour and machine time will eventually be costed).

Required:

(a)   Draw a system flowchart to model the system described.

(b)   Document the Figures 5.19 to 5.22 using an appropriate Document Description Form.

(c)   Make a list of questions you would like to ask at subsequent interviews at City Dyers Ltd.

4   The following data was obtained by interviewing the personnel in the departments concerned.

The design draughtsman creates a design drawing. When finished he removes from file the Part Number Register and examines it for the last used part number. The draughtsman enters the next part number and part description in the register and the part number on the drawing and refiles the Part Number Register.

The design draughtsman passes the design drawing to his Section Leader who checks for any errors. If he finds an error he marks it in red pencil and returns it to the draughtsman, who makes the necessary correction and resubmits the drawing to the Section Leader. The Section Leader rechecks the drawing and passes it to the Chief Designer for signature. If the drawing is correct when first checked the Section Leader passes it straight to the Chief Designer. The Chief Designer signs the drawing in the space provided and passes it to the Print Room.

The Print Room prints two copies of the drawing on a dyeline printer and, placing these together, passes them to the Design Office. The original design drawing is returned separately to the Design Office where it is filed in the 'Master' file.

On reaching the Design Office the No. 1 print is stamped 'Advance Issue' by the draughtsman who then separates the copies, files the No. 2 copy in the Design Office and passes the other to the Technical Engineer. The Technical Engineer checks to see if the drawing is the last for the assembly, if not he files it in a 'Pending' file. If it is the last he withdraws from a 'Pending' file the remaining prints for the assembly, places them together and creates a Parts List (handwritten).

The Technical Engineer, having raised the Parts List, sends it to the machine room supervisor who passes it to the punch operator. The punch operator punches a Breakdown Pack from the Parts List, then machine-sorts the pack into part number order within assembly level. The Parts List is returned to the Technical Engineer (Design Office) who files it.

Having sorted the cards the punch operator uses a tabulator to produce a complete Breakdown Parts List. The cards are filed in the machine room and the Parts List passed to the Technical Engineer who then withdraws a complete set of No. 2 drawing prints from the file, places the whole lot together and passes them to the Process Planner.

The personnel working within the Design Office are:

Draughtsmen

Section Leaders

Chief Designer

Technical Engineers

Required: Draw a system flowchart of these procedures.

6 Logical Modelling: Data Flow Diagrams

6.1 INTRODUCTION

The previous two chapters have examined ways of modelling the current operational systems of the organisation. This knowledge is very important, but these models alone do not provide a sound basis for moving into the design of a replacement. They concentrate on how operations are currently organised and arranged and these structures may not be appropriate for the planned replacement. For example, there is no reason why administrative structures should necessarily survive the transition from manual to computerised systems. The fact that many do is probably testament to the reluctance or inability of organisations to grasp the full strategic possibilities of computerisation. Mimicking current operations tends to lead to 'computerised manual systems' whose very scope limits their chance of success. The point is put most eloquently by Stafford Beer (Beer, 1964).

> The departments or sections of a firm ... are there and are identifiable because of the limited capacity of the human brain. The methods they use and the very tasks they undertake exist in this form because of the limitations of eye, hand and the capacity for human communication ... Our systems are tailored to these limits. What idiocy then to slap these limits on a computer, a machine devised by the wit of man precisely to circumvent them. What anthropomorphism to cut down these machines to the size of human frailty, and to enshrine the inadequacies of men in steel and wire and semiconductors.

System flowcharts are a useful way of showing the passage of documents and data through a system but their very construction emphasises the organisational rather than logical arrangements for processing data. The columnar format shows the different sections, departments, factories or personnel as separate columns. There is a high probability that these will be unwittingly built into the replacement system without questioning their fundamental usefulness to the data handling of the organisation. The administrative arrangements unconsciously become cast in tablets of stone.

Therefore, the Analyst needs tools which separate the logical tasks of an organisation from their administrative trappings. Two important logical tools are covered in this and the succeeding chapter – data flow diagrams and entity-relationship modelling, both supported by a data dictionary. These tools permit the development of a logical model that shows the information processing requirements of an organisation stripped of their current physical arrangements.

The underpinning nature of this logical model is shown in the following diagram.

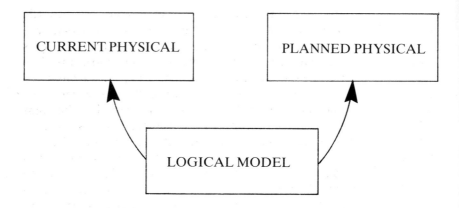

The logical model of the organisation's data requirements remains relatively static whilst the physical reflection of those needs can be changed to exploit new software and hardware opportunities. It is widely acknowledged that many implementations have been too conscious of physical design constraints and have consequently built the limitations of the target software into the system requirements. The existence of an underlying logical model permits such compromises to be explicitly recognised and so possibilities of improving the physical implementation (by using, for example, a new software product) can be recognised more quickly.

Thus two fundamental models of the organisation are suggested. The first is a logical view of the company's information needs which presents a model independent of physical operations. The second is an actual implementation of those requirements which may not fulfill them completely, efficiently or effectively. The latter model is periodically reviewed against new hardware and software opportunities and recommendations made to reduce the gap between the two models.

This chapter introduces a tool for examining how data is transformed and manipulated as it passes through an organisation. Data flow diagrams show the passage of data through a system irrespective of current departmental structures and administrative control procedures. In this way they focus attention upon the logical events required by the system and not the present arrangements for executing those events.

6.2 DATA FLOW DIAGRAMS

Data flow diagramming was introduced in two seminal books (de Marco, 1979 and Gane and Sarson, 1979). Data flow diagrams are central to most Structured Systems Analysis and Design methodologies and the notation is usually similar in concept but differing in shape. This text adopts the NCC symbols (NCC, 1987) which have the advantage of giving more physical space in which to write the associated text. Whatever notation is used it must be stressed that data flow diagrams are graphical, with the guiding principle that a picture is worth a thousand words.It is also important to stress that, in this text, the data flow diagrams (DFD) are being used only for describing the logical model of the system. Many methodologies also use the DFD notation to show the current physical activities. However, there are already adequate models for doing this (such as the system flowchart) and using DFDs for this purpose is unnecessarily constricting and confusing. It can be argued that making the step from the current physical system to the logical model is easier if a different modelling technique is used.

6.2.1 Data Flow Diagrams: Notation and Construction

The data flow diagram depicts the passage of data through a system by using four basic symbols. Each of these may be considered in turn:

Data Flows

A data flow is a route which enables packets of data to travel from one point to another. The flow can be viewed as a road with busloads of data passing along it at certain intervals. Data may flow from a source to a

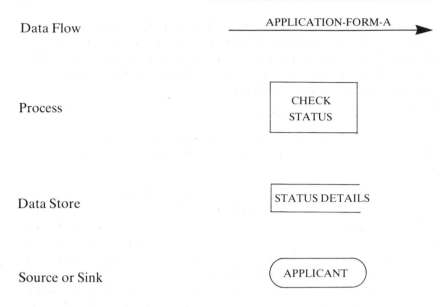

process, or to and from a data store or process. The flow is shown as an arrowed line with the arrowhead showing the direction of flow. In the example given below Part A of an Application Form passes from the applicant (a source) to a process which checks the applicant's status.

Data flows must be named, preferably with titles which clearly describe the flows, and no two data flows should have the same name. If two flows, separated by a process, appear to be identical then it is likely that the process that separates them is doing nothing and should be discarded. The data flows moving in and out of stores do not require names, the store name being sufficient to describe them.

Processes

Processes represent transformations, changing incoming data flows into outgoing data flows. Processes must also be named using descriptions which convey an impression of what happens to the data as it passes through the process. Gane and Sarson suggest that the ideal naming convention is an active verb (extract, compute, verify) followed by an object or object clause. For example:

Verify that customer is credit-worthy.

Other active unambiguous verbs include create, produce, retrieve, store, determine and calculate. Processes may also be numbered to aid identification.

No physical activities or arrangements should intrude into the process description. "Sort records into alphabetical sequence" is not a logical function. Similarly, "Accounts Clerk extracts Supplier Code" is also not allowed because it represents a physical commitment to how the operation will be carried out. Descriptions of such procedures or activities have their place in complementary models, such as the system flowchart, but not in the data flow diagram. In the process shown below (Check Status) an incoming flow – Application-Form-A – is transformed into two outgoing flows, a valid application form (Valid-Appl-Form-A) and a flow which recognises that the form requires further clarification. This latter flow, designed to ensure that the applicant can maintain himself whilst on the course, returns to seek confirmation of financial status and support (Fin-Status-Query).

Data Stores

The store is a repository of data. It may be a card index, a database file or a wastepaper basket. The physical representation is again irrelevant, it is the logical requirement that is important. Stores should also be given convenient descriptive names. Data stores may be included more than once to simplify the presentation of the data flow diagram.

A store may be used in the checking of data. For example, in the diagram shown below the process Check Status requires access to data which permits this checking. The data items required to correctly carry out this process must be available in the store Status Details. The arrow is single headed and points towards the process. This is to signify that the process does not alter the contents of the store, it uses only the data available. However, if the contents of the store are altered by the process, as well as being read, then the diagram uses a double-headed arrow.

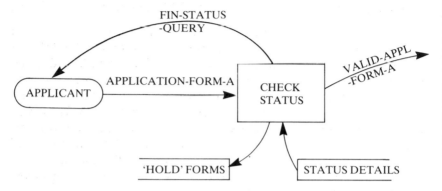

In this way a single-headed arrow shows READ (looking at the data only) or WRITE (changing data only) operations. The creation of the store 'Hold' Forms is a Write-Only function of the Check Status process. In circumstances where the data is both examined and changed (READ and WRITE) then a double-headed arrow is used.

Sources or Sinks

A source or sink is a person or part of an organisation which enters or receives data from the system but is considered to be outside the context of the data flow model. The definition of these external entities may change as analysis proceeds. Entities first considered to be outside the system may be incorporated into the model as the project progresses, and vice versa. The source or sink may be duplicated in a completed data flow diagram to permit simpler presentation. Both sources and sinks force the consideration of the boundary of the system as they are viewed as external to the system under investigation. It must be recognised that subsequent analysis may demand that certain sources or sinks are brought into the detailed system consideration.

6.2.2 Levelling the Data Flow Diagram

A good modelling technique should have an easy to use hierarchy. A book of road maps clearly illustrates this. In planning a route, the traveller looks at the front of the atlas at a page that shows the major cities and page numbers where more details can be found. Turning to the relevant page, he finds the major trunk roads with towns represented as shaded areas and hence can chose a suitable route. At the back of the atlas are often found detailed street plans which enable him to pinpoint his destination. The maps have different scales for different purposes and have an understandable hierarchy.

Information systems models also need such a hierarchy. The data flow diagram has a simple consistent way of representing the different levels. Each process is exploded into a lower level data flow diagram until the specification of the process can be presented on an A4 sized data dictionary entry (see later). Thus it is possible to present a series of data flow diagrams representing increasing levels of detail. These may then be used with different levels of user staff. The high level overview model may be discussed with senior management, whilst more detailed charts are the focus of discussions with appropriate line managers. The convention of calling the highest level diagram Level 0 has been adopted with subsequent levels designated 1, 2, etc.

The numbering system of the Level 0 model needs to be extended to the lower level diagrams so that they can easily be referenced back to their 'parent' processes. This is achieved by using a decimal numbering system. Thus process 3 may be decomposed into 3.1, 3.2, 3.3, etc and, if a further level is required, 3.1 into 3.1.1, 3.1.2, 3.1.3, etc. Figure 6.1 demonstrates levelling for the example introduced in the previous section. The diagram shows how the process Check Status actually comprises three individual checks – determining nationality status, producing examination equivalents and checking financial support.

It is important to check that all flows and stores accessed by the process in the higher level box are actually used in the lower level decomposition. Data flow diagrams should be balanced so that data flows in and out of a process must appear on the data flow diagram that is a decomposition of that process. If a flow does not appear then the reason for its existence at the higher level must be closely examined. Similarly, flows should not be produced that do not exist in the parent process. If such a flow seems to be necessary then all higher level diagrams should be altered so that the whole model remains balanced.

6.2.3 Hints for Drawing Data Flow Diagrams

Do not be worried by detail. Tom de Marco quotes Gerry Weinberg's

First Level Diagram (Extract)

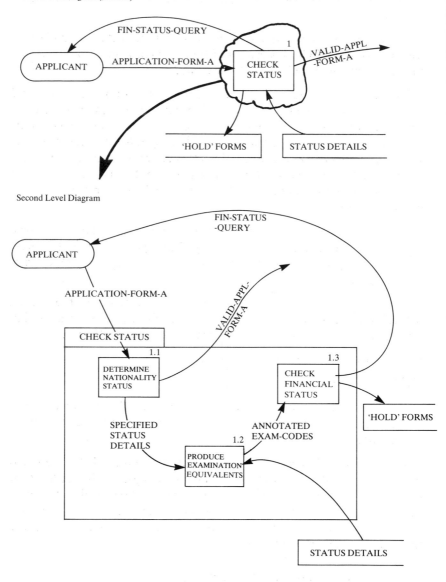

Second Level Diagram

Figure 6.1   Explosion of a Process into a lower level Data Flow Diagram

Lump Law: "If we want to learn anything, we musn't try to learn everything".

Try to identify sinks and sources. This will provide a boundary in which to work and also gives a set of flows – what is the source sending to the system and what is the sink receiving?

Start with an obvious input from a source, or an output to a sink and concentrate on the data flows. Place a box where a process is required to transform one flow into another. These boxes can be named later. Consider whether there is enough data flowing into the process (from stores and flows) to perform the stated transformation and to produce the required output. This self-checking nature of data flow diagrams can help identify extra data items that were not obvious from a narrative or verbal description.

The first data flow diagram that you draw of a system is bound to be a tangled scribble and is probably wrong. Start again and modify. Gane and Sarson suggest that at least three drafts of the higher level diagram will be needed.

Pay no attention to timing considerations except for the logical precedence of activities. It is often useful to follow a typical transaction through the system modelling the logical activities that affect it.

Most processes access a store of some kind.

Do not use a template or a pen in the first draft.

Do not worry about absolute correctness. Even wrong (not dramatically wrong!) diagrams have much to contribute to understanding.

The data flow diagram is a useful and immediate modelling tool and its limited notation is easy to handle and unambiguous. But, although it provides a clear impression of how data passes through the system, it fails to represent that passage in sufficient detail for the subsequent design stage. Consequently a supporting model is required to capture that detail. A data dictionary is suggested.

6.3   DATA DICTIONARY: DATA FLOW

A data dictionary is simply a record of data about data. It may be manually compiled (on A4 sized sheets in a loose-leaf folder) or it may be a fully automated package. There are considerable benefits to be gained from using an automated dictionary and these will be discussed in the final chapter. However, for the present purpose of illustration, a simple manual system will be assumed.

In structured systems analysis and design it is usual to hold data about three of the four main constituents of the data flow diagram. These are the store, the process and the flow. It is also necessary to record information about two further fundamental concepts that do not enter explicitly into the diagram. The first of these is the *data element.* This is an item of data which it is not meaningful to decompose any further for the task in hand. These elements may be viewed as fundamental building blocks of the system. Typical examples might be Customer-Name, Invoice-Date and Net-Price. It must be recognised that these elements are seldom absolute. For example, a data element in one system Customer-Code may be two or three elements in another system – Cust-Loc-Code, Cust-Pay-Code, etc. Hence the importance of recognising that data elements are defined within the requirements of the 'task at hand'.

The second new concept is that of the *data structure.* This may be viewed as a collection of data elements that regularly appear together. Thus a data structure can be defined and then used as a shorthand reference in flow, process and store definition. In fact, the data structure can be interpreted as the most basic part of the whole diagramming technique, as data flows are simply structures in motion and stores structures at rest.

This gives the following hierarchy:

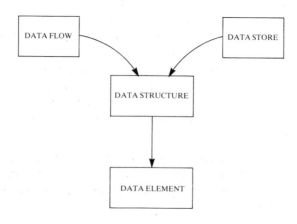

It is necessary to hold data dictionary entries about data elements, data structures, data flows, data stores and processes. The structure of the data dictionary for each of these will vary and Figures 6.2–6.6, whose format is largely based on those of Gane and Sarson, are meant to be illustrative. Chapter 8 examines the issue of content in more detail.

The structure of the data dictionary entries will now be examined in more detail.

Data Dictionary Entry: Data Element (Figure 6.2)

Name: A meaningful unique name.

Description: A short description of the meaning of the data element. An example might be included.

Aliases: Several departments may refer to the element by a different name or term. This has to be explicitly recognised and may require a large amount of detective work. Different Officers in an examination system used the terms 'unit' and 'module' to refer to the same data element. This was recognised only through the analysis work required in the compilation of a data dictionary.

Type: Usually Character, Numeric or Alphanumeric.

Format: To prepare for format checks in the subsequent system design. A convention of representing numbers by nines (9) and characters by Xs can be adopted.

Values: Discrete data elements have a meaning associated with each value. A good example of this is the rating code system in the United Kingdom where 100 is a House, 101 a House and Garden through to 700 for a Warehouse with Solarium. Such figures may also be viewed as having a continuous range, any values of the data element outside the values 100–700 are incorrect and need rejecting. Data about ranges and discrete values provides the basis of many of the data validation checks required in the subsequent system.

Security: Who (or which level of employee) is allowed to modify, add or delete a given data item. This will be important in the design of security features such as passwords and audit checks.

DATA ELEMENT NAME: *APPL-REF*

SHORT DESCRIPTION *A unique code given by the college to each Applicant applying for a full-time course (Application-Reference)*

ALIASES *Applicant-number*   TYPE *Alphanumeric*

FORMAT *XX99999*

VALUES

DISCRETE	CONTINUOUS
Not applicable	*LP00001-LP99999 {this year}* *XX00001-XX99999 (general)* *Note: unlikely for values to exceed LPZ0000*

SECURITY

Created only by admissions tutor
Deleted only by admissions tutor
May be viewed by all staff members

EDITING

Not Applicable

COMMENTS

Serial number. Two character prefix changes from year to year

Figure 6.2   Data Dictionary: Data Element

Editing: This may concern the way in which the data is produced from the system. For example, should a credit of £30 be shown as −30, +30, 30CR or (30)?

Comments: A final section in which to record special information about this data element.

Data Dictionary Entry: Data Structure (Figure 6.3)

A data structure is made up of data elements and other data structures. Thus the dictionary entry contains a list of associated elements and structures which are documented elsewhere in the data dictionary. The list may also contain data elements which are either optional, repeated or mutually exclusive. The following convention can be adopted:

Optional Structure: Placed in square brackets.

[PREV-SURNAME]

Alternate Structure: Placed in braces.

$$\left\{ \begin{array}{l} \text{PARENT-NAME} \\ \text{GUARDIAN-NAME} \end{array} \right\}$$

Iterations of structure: Marked with an asterisk.

COURSE-APPLIED* (1–5)

With the number of iterations placed, if known, in parentheses. In this case an Applicant can apply for 1,2,3,4 or 5 courses – but no more.

Volume information collected at the end of the form will be needed when the size of the system is estimated.

Data Dictionary Entry: Data Store (Figure 6.4)

The usefulness of the data structure can now be seen in the description of the data store. The content of the store can be described much more economically and with less chance of error. Furthermore, the interrelationships of parts of the system are also represented by the occurrences of the structures. It may also be possible to include the Search criteria in the definition. An example is given illustrating how the request Applicant-Status-Query (the current status of the Applicant's application) is always searched for on Appl-Ref – the reference given to him or her by the College.

DATA STRUCTURE NAME: APPLICANT

SHORT DESCRIPTION Describes data elements

associated with a person applying for a course
place

CONTENT APPLICANT :

APPL-REF	SEX
SURNAME	MARITAL-STATUS
FORENAME	BIRTH-DATE
DESIGNATION	BIRTH-COUNTRY
[PREV-SURNAME]	BIRTH-RESIDENCE
HOME-ADDRESS	NATIONALITY
[CORR-ADDRESS]	PARENT-NAME
[HOME-TEL NO]	GUARDIAN-NAME
[CORR-TEL NO]	COURSE APPLIED * (1-5)

VOLUME INFORMATION

2000 / YEAR

COMMENTS

Figure 6.3   Data Dictionary: Data Structure

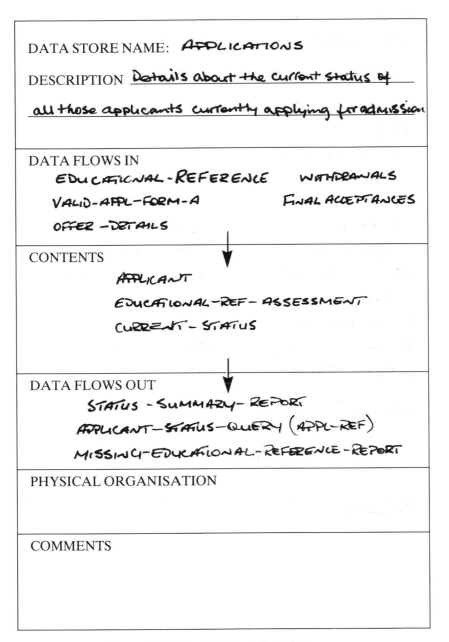

DATA STORE NAME: APPLICATIONS

DESCRIPTION Details about the current status of all those applicants currently applying for admission.

DATA FLOWS IN

EDUCATIONAL-REFERENCE   WITHDRAWALS

VALID-APPL-FORM-A   FINAL ACCEPTANCES

OFFER-DETAILS

CONTENTS

APPLICANT

EDUCATIONAL-REF-ASSESSMENT

CURRENT-STATUS

DATA FLOWS OUT

STATUS-SUMMARY-REPORT

APPLICANT-STATUS-QUERY (APPL-REF)

MISSING-EDUCATIONAL-REFERENCE-REPORT

PHYSICAL ORGANISATION

COMMENTS

Figure 6.4   Data Dictionary: Data Store

The dictionary entry for the data store will assist subsequent definition and organisation of files.

Data Dictionary Entry: Data Flow (Figure 6.5)

Data flows are again defined in terms of data structures. The source and destination of each flow will be given using the reference numbers allocated to the processes in the data flow diagram. Volume information will assist in system sizing as well as in the calculation of input and output times.

Data flow entries define input and output requirements. They should determine the content of forms and reports as well as guiding the technology used to capture or produce them.

Data Dictionary Entry: Process (Figure 6.6)

The logic of a process may be documented with a variety of tools, such as decision tables, decision trees and structured English. Some process definition tools have already been introduced in the previous chapter, whilst others are examined in the companion text *Introducing Systems Design*. Inputs and Outputs again contain data structures or elements which have their own dictionary entries as well as being referenced by the stores and flows used and documented in the process.

The Process entries in the data dictionary form the basis of clerical procedure documents, computer programs or some combination of the two.

In summary, the data dictionary represents an important corporate resource which can be used as a foundation for the logical system definition. Three points are worth stressing:

— The dictionary is not a static mechanism. The information in it will be built up over a period of time as the Analyst gains greater understanding through fact finding.

— The data dictionary, in conjunction with the data flow model, provides an important part of the logical bridge between analysis and design.

— The data dictionary has to support a complex web of inter-relationships. A single data element may appear in many data structures, data flows, data stores and processes. The effect of amending a data element has to be 'traced' through the system to identify possible problems and side-effects. The automatic mainte-

DATA FLOW NAME: APPLICATION-FORM-A

SOURCE REF: Source DESCRIPTION APPLICANT

DESTN REF: 1 DESCRIPTION CHECK STATUS

DATA FLOW DESCRIPTION Describes an application

form submitted by an applicant

CONTENT:

COURSE

APPLICANT

APPLICANT-EDUCATION

APPLICANT-SPECIAL-NEEDS

APPLICANT-PAYMENT-METHOD

REFEREE

VOLUME INFORMATION

2000/year

COMMENTS APPLICATION FORM-B IS The
EDUCATIONAL REFERENCE. This will accompany
PART A if the flow occurs before the 16th July

Figure 6.5 Data Dictionary: Data Flow

PROCESS NAME: 1.1 Determine Nationality Status   DESCRIPTION: To determine whether student is classed as Specified (overseas)

INPUTS	LOGIC	OUTPUT
APPLICATION-FORM-A	If BIRTH-COUNTRY = "ENGLAND" or "SCOTLAND" or "NORTHERN IRELAND" or "WALES" or BIRTH-RESIDENCE = "UNITED KINGDOM" AND NATIONALITY = "BRITISH" THEN Mark the Application form "Valid" and Pass to relevant Admissions tutor ELSE Mark the Application form "Specified" and undertake Examination checks ENDIF	VALID-APPL-FORM-A SPECIFIED-STATUS -DETAILS

COMMENTS Specified Students must be subject to examination equivalence checks (1.2) and stringent financial verification (1.3)

REFERENCE (FULL LOGIC DESCRIPTION)

Figure 6.6   Data Dictionary: Process

nance of these links is clearly necessary in all but the most trivial of applications and hence the importance of data dictionary software.

6.4 DATA FLOW AT INFOSYS

One of the features of the data flow diagram is that it can be used for both top-down and bottom-up design. Figure 6.7 shows the system flowchart introduced in the previous chapter (Figure 5.5) redrawn as a data flow diagram. In this example, the DFD is being used as a bottom-up tool, producing a logical version of the current operational system. Notice how all physical references have been stripped away, leaving only the tasks that are logically needed by the system.

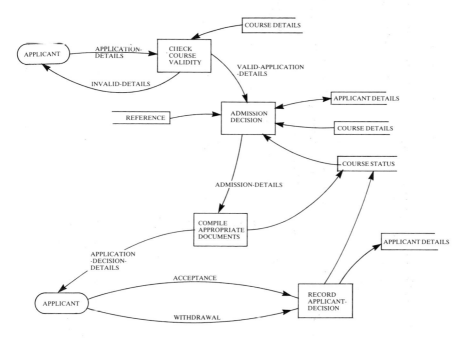

Figure 6.7   DFD of the System Flowchart shown in 5.5

Data flow diagrams are also useful in top-down design. This can be illustrated by returning to InfoSys and examining one of the issues raised in the Strategic Plan (Chapter 2). You will recall that it was agreed that the company should start a Mail Order Book Club. This club should offer texts from the InfoSys catalogue at reduced prices along with other significant computer books published by rival companies. It was envisaged that the Book Club would run along the lines of other specialised clubs, with introductory special offers and a requirement for each member to purchase no fewer than four books in the first year.

The Strategic Plan made it quite clear that the venture was to be supported by an efficient information system that would help minimise staffing as well as providing detailed statistics on the success of the Club. The subsequent Feasibility Study (see Chapter 3) resulted in the project being given the go-ahead and so detailed analysis work commenced.

The absence of a current operational system guided the project team towards logical modelling tools. A number of important transactions and rules were identified.

1   That all orders will be cash, cheque or credit card with order and that no direct credit facilities are to be offered. This simplifies the cash flow of the Book Club as well as eliminating the problems of debt collection. All cheque and credit card transactions will be cleared before the order is despatched.

2   The system has to be linked through to order despatch. Hence all orders received in one day should be consolidated and a picking list generated for the stores clerks.

3   Stock levels must be maintained and monitored. Books which have reached their reorder level should be ordered from the InfoSys Publications department or the appropriate publisher. If the text is not in print then it will be deleted from the Club's list. The stock levels will also be monitored for slow-moving stock that can be made into Special Offers.

4   Details of all customers will be maintained and used as a mailshot source for both the Book Club and the rest of the company. Mailshots will be made on a selective basis. For example: "Send details of our forthcoming Database Course to all those Book Club members and past members who have ordered a database book in the last three years".

5   All order and mailshot labels will be produced automatically.

6   A comprehensive reporting system is required that summarises both the activities of customers as well as the club as a whole.

7   The club must react to order queries made by customers.

In an attempt to show the stages of developing a DFD, a draft diagram for the Book Club operation is shown in Figure 6.8 and a Level 0 data flow diagram in Figure 6.9. Figure 6.10 explodes one of the processes into a lower level diagram.

Figure 6.8 is an attempt at getting to grips with the essential features of the proposed Book Club operation. In some respects it provides a stage between a Rich Picture (see Chapter 2) and a Level 0 data flow diagram. The notation of the data flow diagram has been used where appropriate to ease the transition into the Level 0 model.

The Level 0 model illustrates a number of features in data flow diagramming. The first concerns the reduction of the payment, customer and order checks into one process called Verify-Order-Details. The model is at its most accessible when it has only six or seven processes. If three of these are quickly taken up with simple validation checks then there is little scope for much activity on the rest of the diagram. Furthermore, the error conditions which are normally associated with validation are also dropped from this level to help clarity. Only one is retained (Uncleared-Payment) and this was actually inserted after the Level 1 diagram for Process 1 was completed to permit balancing.

The External Publisher has been dropped from the diagram. This is because the Publications department effectively acts like an outside company (sending an invoice, delivery note, etc) and so can represent all order movements.

No distinction has been made between special mailshots and the monthly catalogue. Both are dealt with under Process 7 (Compile Mailshot) and it is assumed that the monthly offer list is just a particular type of mailshot.

Two further processes have been added. Process 8: Evaluate Mailshot. This is an important feature of the project. The Book Club must target its potential customers very carefully. Thus the success of all mailshots must be analysed. Process 9: Respond-To-Order-Query. Customers must be able to make enquiries about their current order.

It must be stressed that the stores are only logical. Whether it is necessary to have a Customers' and an Orders' store is not an issue at

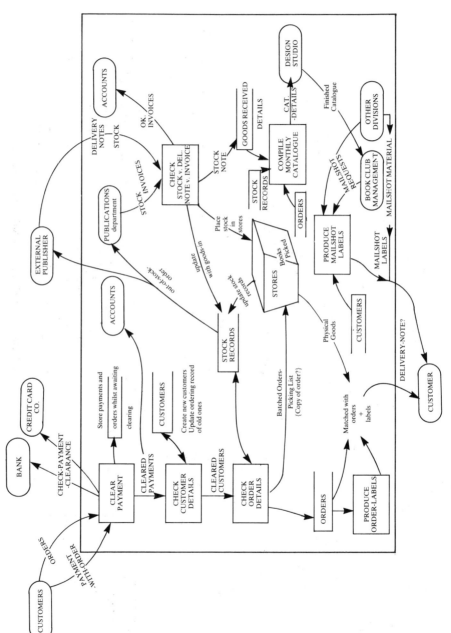

Figure 6.8   A Rough Diagram for the Book Club Project

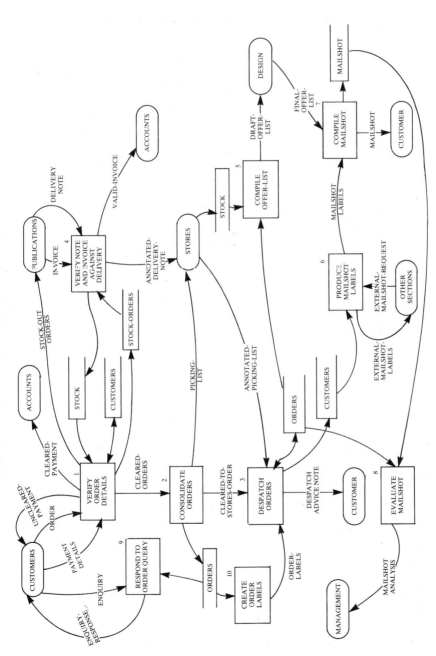

Figure 6.9   Level 0 Data Flow Diagram: Book Club Project

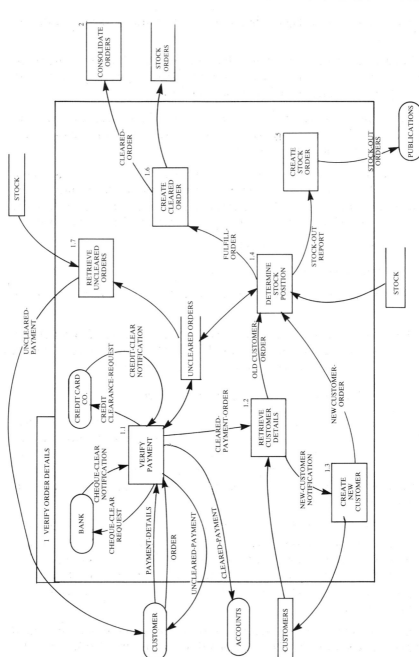

Figure 6.10   Level 1 Data Flow Diagram: Book Club Project

present. They have logical significance in permitting the processes to take place.

Finally, it should be clear that 10 processes push the DFD close to its limits of clarity. The diagram might benefit from a (fourth!) redraft and perhaps the combination of processes 6 and 10.

The data flow diagram on the facing page (Figure 6.10) provides the explosion of Process 1. The payment checks with the bank and the credit card companies have now been included.

An Uncleared Orders store has been established to hold all those orders which cannot be currently fulfilled. This would include orders which have not cleared the payment checks or for which stock is not currently available. An uncleared payment is raised for all orders that:

— do not clear payment checks;

— do not enclose the correct payment;

— have remained in the Uncleared Orders store for more than 28 days.

This latter rule is necessary because the Club has agreed to conditions of '28 days delivery or money refunded'. Orders which have remained uncleared for that length of time are picked up by Process 1.7.

6.5   DATA FLOW: FINAL CONSIDERATIONS

In a recent study (Sumner and Sitek, 1986) it was recorded that data flow diagrams were being employed in 15 of the 45 projects in the survey, compared with 26 recording the use of systems flowcharts. This made it the fourth most popular requirements analysis tool after interviews (40 projects) and data dictionaries (26 projects). However, the authors concluded that:

> . . . although most of the respondents acknowledged the benefits of using structured tools in requirements analysis and design, these tools were not being widely used in actual systems development projects, largely because of their lack of acceptance by data processing staff, and the fact that they were perceived as time-consuming to use.

Edward Yourdon has written pessimistically about the structured analysis revolution that he contributed to (Yourdon, 1986). He writes that:

Today, only about 10% of the DP organisations in North America practise structured techniques in a disciplined fashion.

The issue of use is of importance to the student reader. It might be legitimately claimed that it is not worth learning these techniques if they are not being widely applied. However, four observations must be made:

1   Inadequate training contributes to the lack of acceptance of structured methodologies. Many DP staff are unsure how and where to apply these new techniques.

2   All ideas take time to filter into everyday use. Yourdon comments that it took the military 75 years to switch from muskets to rifles.

3   The manual maintenance of data flow diagrams is very time-consuming. The development of computer-based design tools is beginning to eradicate this problem. CAD/CAM tools have changed the way that most designers have worked in all engineering disciplines. It is now the turn of computer systems designers. Data flow diagrammers permit the computer-based design, amendment and validation of models.

4   Finally, we believe that the acceptance of data flow diagrams has been held back by their use in modelling the current physical system. This makes it difficult for developers to move from the physical trappings of the present operations to their logical equivalents. Hence, data flow diagrams are used in this text only for logical modelling.

6.6   SUMMARY

This chapter has:

—   Restated the need for logical modelling.

—   Introduced the notation and construction of a powerful logical model – the data flow diagram.

—   Demonstrated the development of a data flow diagram in the context of the Case Study.

—   Illustrated the content and structure of a data dictionary to support the data flow model.

—   Concluded with a brief assessment of the data flow diagram, suggesting factors that affect its use and acceptance.

6.7 PRACTICAL EXERCISES AND DISCUSSION POINTS

1 Produce a data flow diagram of the order handling narrative given in the Practical Exercises at the end of Chapter 5.

2 Models have a number of uses. Compare the data flow diagram produced in question 1 with the system flowchart developed at the end of Chapter 5. Comment on:

Their relative ease of construction: How long did it take to produce each model?

How many redrafts were necessary?

How good is each model at communicating its features? Give the models to a colleague and ask him or her to assess their clarity and meaning.

What is each model contributing to the Analyst's overall task?

What are the overall strengths and weaknesses of each model?

3 Construct data dictionary entries for the flows, processes and stores that contribute to the process Verify Payment (1.1) in Figure 6.10. Make any assumptions that you feel are required. The documentation should include the following:

Data Flows

　　Cheque-Clear-Request
　　Cheque-Clear-Notification
　　Credit-Clearance-Request
　　Credit-Clear-Notification
　　Payment-Details
　　Order
　　Uncleared-Payment
　　Cleared-Payment-Order
　　Cleared-Payment

Process

　　Verify Payment

Data Stores

　　Uncleared Orders

PLUS all data elements and structures used in these processes, stores and flows.

- You may feel it is necessary to create a Level 2 diagram.

- This exercise should give you an insight into the potential size and compilation time of a manual data dictionary!

4   Investigate three different structured methodologies and compare the nomenclature and notation of the data flow diagrams (or their equivalent) in each. What features do any of them have which you feel might improve the basic notation introduced in this chapter?

5   Undertake a market survey of the data flow diagramming tools currently available. Undertake an evaluation of one tool (many companies offer cheap demonstration versions) and assess its usefulness in model development.

7 Logical Modelling: Data Analysis

7.1 INTRODUCTION

The data flow diagrams introduced in the previous chapter are primarily concerned with the logical modelling of the processes required in the system. This chapter now looks at the complementary technique of data analysis which the Analyst can use to help him examine, understand and model the logical structure of the data. The aim of data analysis is to build a data model that *supports* but is not *driven* by the identified processes. The two techniques are aimed at producing complementary logical views of the system requirements.

Data is the raw building block of all information systems and the objective of data analysis is to express the data structure in a logical, concise and useful way. It has been pointed out that this is a time-consuming but necessary activity which can easily be justified because information is a valuable corporate resource (Veryard, 1984). The organisation needs a clear and accurate knowledge of the data structures underlying its information requirements, whether they are computerised or not.

Considerable benefits accrue from such a detailed analysis of the data structures independently of the physical design associated with any hardware and software implementation. The Analyst has much to gain by taking this wider and longer term view of the organisation's data. For example:

— The data structure tends to be more stable than a user's functional requirements or organisational structure. A business may change over time in how it deals with data and the information required from it, but the underlying structure of this data is unlikely to change as much.

— Users often find it easier to understand and criticise data models which reflect the actual data structure of the organisation rather than other models used in system specification.

169

— A lack of understanding of this logical data structure often leads to poorly designed, troublesome and inflexible systems. Such systems may be quicker to implement but may underperform and be costly to maintain in the long term.

— It promotes the design and development of shared data systems with the benefit of greater control over data security and integrity and less duplication of stored data.

This chapter takes a top-down view of the logical structure of data required to support the information systems needed by the organisation. Two complementary tools for data analysis are discussed: *Entity-Relationship modelling* and *Normalisation* of data tables. The use of these techniques is further illustrated with examples from the InfoSys case study. The long-term advantages for designing flexible information systems will be discussed in depth in the companion book *Introducing Systems Design*.

7.2   ENTITY-RELATIONSHIP MODELLING

Entity-relationship modelling is a technique for analysing and modelling the organisation's data requirements. It requires the discovery of the data elements needed to support the information systems and represents their structure in a clear, concise diagram.

7.2.1   Concepts

Entity-relationship modelling uses three basic concepts, *entities*, their *attributes* and the *relationships* which exist between entities.

The Entity

This is the term used to describe something that the enterprise recognises in the area under investigation and wishes to collect and store data about. An entity may vary from a physical object, such as a machine, to a more abstract concept such as a sales area. In all instances the entity should be capable of being uniquely identified. In the example introduced in the previous chapter, Book, Member and Order are suggested as likely entities in the model of the Mail Order Book Club.

Attributes

Entities have attributes. These are the data elements which have been collected and stored in the data dictionary. For example, Member may have the attributes member-name, member-number, member-address, etc, while the attributes of Book might include title, author and publisher.

Relationship

Entities are linked together by the concept of relationship. So the entity Member may be associated with the entity Order by a relationship that can be called, say, Places.

These three concepts form the basis of entity-relationship modelling and this will now be examined in detail. However, it is useful to clarify a few points before embarking upon a technical exposition:

1   It is very difficult to perform entity-relationship modelling in the early stages of problem understanding and definition. It needs to be preceded by the process modelling work that has been described in the previous two chapters. Good problem understanding must come before good entity-relationship modelling.

2   There are no 'absolute' entities and attributes. The importance of a particular 'thing' will vary with the nature and scope of the system under consideration. For example, Book may be identified as an entity in the Book Club because it has evident importance to the enterprise. But in another context, book may appear as an attribute of an entity. For example, in a personnel records system for the consultancy staff of InfoSys, book may be an attribute of the entity Employee, used to record published works for an Employee entity.

An entity normally needs more than one attribute to be an entity. For example, a child may be an attribute of an Employee entity, but if other attributes relating to the child (say name, date-of-birth, and sex) are required it is better to introduce a new entity rather than allocate the three attributes to Employee (child-name, child-date-of-birth, child-sex). Thus the context of the model is very important – one system's attribute may be another's entity – because entity-relationship modelling is not a substitute for the perspective of common sense.

3   It is important to distinguish between the *entity type* and the *entity occurrence*. The entity type is Book, whilst the entity occurrence is a specific example of Book, such as 'Data Analysis for Data Base

Design' by DR Howe, or 'Systems Analysis' by A Parkin. The same distinction can be used when considering relationships. There is one *relationship type* Places, but there are likely to be hundreds of *relationship occurrences* representing specific orders placed by individual members. In this chapter the terms 'entity' and 'relationship' will be used to refer to entity and relationship types.

Identifiers

The proviso, previously mentioned, that the entity should be capable of being uniquely identified, is very important. A particular entity occurrence should be recognised by the values of a particular attribute or combination of attributes. For example, a member may be identified by the value of attribute member-name or a specific book recognised from the combination of two attribute values author and title. This identifying attribute or combination of attributes is termed the *entity identifier*.

The choice of the entity identifier must be guided by its ability to uniquely identify an entity occurrence. For example, member-number may be preferred to member-name because the organisation may wish to allow more than one 'Smith J' to be a member of the Book Club. (Similarly, author and title may not be acceptable for the entity Book because it is possible that two books will have both titles and authors that coincided.) This is not to say that such an identifier will always be unusable. In a small Book Club with a limited number of members it is possible that it will be acceptable to 'fiddle' members' names in the rare cases where the identifier would otherwise fail, for example by referring to Smith J/1, Smith J/2.

But in a larger system the Analyst will normally need to introduce an attribute to guarantee uniqueness. Books are given a unique International Standard Book Number (ISB number or ISBN) so there is a powerful argument for including this attribute in the entity definition and for making it the entity identifier. Other subordinate factors that influence the choice of identifier are stability (member-number is more stable than member-name, member-address) and brevity (member-number takes up less physical space than member-name, member-address).

Relationship occurrences may often be uniquely identified by the identifiers of the entities participating in the relationship. For example, the occurrence of a particular book on a particular order may be identified by the relationship identifier (ISBN order-number) where these two attributes are the respective identifiers of the two entities Book and Order. But there will be other instances where just one of the entity identifiers will suffice as well as occasions which demand additional attributes to give a unique identifier.

7.2.2 Entity-relationship Diagrams

Several areas of understanding will contribute to the Analyst's recognition of important entities involved in a particular information system. Candidate entities will have emerged through the Strategic Planning, Feasibility Study, and Fact Finding and Recording stages. Further details of the actual data elements required to support the information systems will be documented in the data dictionary, while details of how these data elements are referred to and transformed by the system processes are modelled by the data flow diagrams. Experienced Analysts are able to use these various models and stages to identify likely entities and relationships. However, inexperienced Analysts often find it difficult to conceive the entities involved and so must seek guidance. A suitable suggestion is to consider the data stores of the data flow diagram as initial candidate entities.

The interaction of entities via relationships can be illustrated by examples of entity occurrences of Members, Orders and Books, and the relationships between them in the form of an entity-relationship occurrence diagram (Figure 7.1).

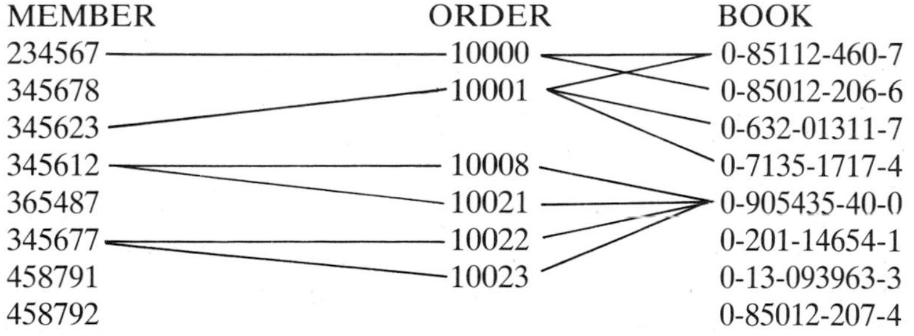

Figure 7.1   Entity-relationship Occurrence Diagram

A more concise picture of the entities and relationships is given by an entity-relationship diagram (Figure 7.2).

Figure 7.2   Entity-relationship Diagram

In the entity-relationship diagram (Figure 7.2), rectangles are used to display entities and relationships arc described within diamond-shaped boxes. (The connecting lines on the entity-relationship occurrence diagram show which entities are associated by each relationship.) The naming of the entities is usually relatively straightforward, but the relationship types often pose more of a problem. It is often difficult to think of a suitable name for a relationship and one may be tempted to call them all 'Has'. Troublesome relationships may be named by a combination of the names of the associated entities (eg Order-Book). This convention also alleviates the difficulty of finding a name which satisfactorily describes the two-way nature of a relationship. For example, the relationship between Order and Book in the entity-relationship diagram (Figure 7.2) could have been called Itemises in the Order to Book direction (ie an Order itemises a Book). However, this is inappropriate in reverse as a Book cannot be said to itemise an Order. In contrast, a relationship name like Order-Book (or Book-Order) is less directional. (Order has an Order-Book relationship with Book; Book has an Order-Book relationship with Order.)

7.2.3   Degree of Relationship

One of the most important properties of a relationship linking entities in a model is the degree of this relationship. In the earlier example two relationships were identified, those between Member and Order and between Order and Book. The relationship occurrence lines shown on the entity-relationship occurrence diagram (Figure 7.1) illustrate two different degrees of relationship.

Consider first the relationship occurrences between Member and Order. These link one occurrence of Member to one or many occurrences of the entity Order. Thus the data model allows a particular member to place one or more orders and this information about the relationship information must be recorded. For example Member 345612 has placed Orders 10008 and 10021. But it will only be possible to record one member for a particular order. For example, Order

10008 was placed only by Member 345612. The model does not support the possibility of an order belonging to more than one member, and so would be incorrect if this were in fact permitted in the actual system under consideration.

The relationship occurrences between Member and Order fan out in one direction only. For relationships of this kind the degree of the relationship is said to be 1:N (or one-to-many).

The 1:N relationship between Member and Order is shown on the lines connecting the Places relationship symbol to the entity symbols on the entity-relationship diagram (Figure 7.3). This notation expresses the existence of the following enterprise rules:

A member may place many orders;

An order may be placed by, at most, one member.

Figure 7.3   Degree of Relationship

Detailed fact finding will enable the Analyst to decide if statements such as these reflect the rules of the organisation. A different degree of relationship is illustrated by the relationship named Order-Book. On the entity-relationship occurrence diagram (Figure 7.1) several relationship occurrences fan out in both directions to link *many* Order entities to *many* Book entities. This demonstrates an M:N relationship (or many-to-many). This type of relationship enables the system to record the fact that an order may itemise several books. For example, Order-number 10001 itemises several books identified as ISBN 0-85112-460-7, 0-632-01311-7 and 0-7135-1717-4. Similarly, a book may be itemised on many orders. For example the Book identified as ISBN 0-85112-460-7 is itemised on Order-numbers 10000 and 10001. In this instance the following enterprise rules apply:

An order may be for many books;

A book may appear on many orders.

The distinction between 1:N and M:N relationships is important in later systems design and so should be carefully considered. For

example, if the Order-Book relationship were to be designated as a
1:N (see Figure 7.4), then the model would allow only a book to be
itemised on one order at a time. Consequently, whenever another
order for a particular book was placed the information of the previous
Order-Book relationship would be lost. Note that the direction of

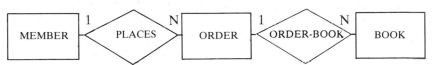

**Figure 7.4   An Entity-relationship Diagram Showing a More Restricted
Order-Book Relationship**

the 1:N relationship is significant. If the direction were reversed, then a
book could be itemised on many orders but each order would be
limited to only one book.

7.2.4   Filling in the Detail

Having identified the initial structure of the entity-relationship model
the Analyst needs to discover other necessary attributes and decide
which entity to assign each one to. Data elements previously identified
in the data dictionary are a major source of attributes. The process of
allocation is likely to highlight some problems in the initial model and so
the Analyst may need to make changes to reflect this. It must be
stressed that model building is an iterative process in which the Analyst
continually reassesses the structure and content of the data model in the
light of his understanding of the organisation's requirements.

Further attributes from the data elements of the Member – Order –
Book example are listed below:

order-date

quantity-ordered

member-name

member-address

order-delivery-address

title

price

publication-date

Some attributes need not be included in the entity-relationship model as they may be derived from the values of other attributes. For example, order-value is calculated from the multiplication and summation of individual values of quantity-ordered and their respective values of price.

Order-date and order-delivery-address may clearly be associated with the Order entity while member-name and member-address can be assigned to the entity Member. Furthermore, the attributes title, price and publication-date logically belong with the entity Book. But a problem arises when considering where the attribute quantity-ordered should be placed. To assign it to the Order entity would be erroneous since there may be several values of quantity-ordered, one for each particular book on the order. It would not be clear which book a quantity-ordered value refers to. Similarly, to assign it to the Book entity is also problematic as there may again be several values of quantity-ordered for a particular book. Which order is being referred to by any one of these values? The situation would, if anything, be even worse if quantity-ordered were to be assigned to the Member entity, as it would not always be possible to tell to which order and book a quantity-ordered value should apply.

The solution to this problem of attribute allocation lies in the decomposition of a many-to-many relationship. This can be done by treating the relationship Order-Book as a separate entity. This may appear to be somewhat arbitrary – after all an entity was defined as a thing identifiable to the enterprise – but it is quite legitimate in the model. Things of importance to the modelling of the application may not be evident to those working in the actual organisation. The name of this new entity may be left as Order-Book but a more appropriate term may emerge as the Analyst examines the identity of the entity in search of a suitable name. In this example Order-Book may be better termed Order-Line since it itemises or identifies the particular part or section of an order relating to one book. Thus quantity-ordered can now be satisfactorily placed in the new entity Order-Line since a single value is associated with the values of the composite identifier order-number and ISBN.

Figures 7.5(a) and (b) show the decomposition of the M:N relationship into two 1:N relationships. The occurrence diagram (Figure 7.5(a)) shows the original M:N relationship. By treating each Order-Book relationship occurrence as an Order-Line entity occurrence the diagram in Figure 7.5(b) is obtained. It is easy to see that the relationship between Order and Order-Line is 1:N and that the relationship between Book and Order-Line is also 1:N.

Figure 7.6 shows the effect of the decomposition illustrated by Figure 7.5(b) on the entity-relationship diagram in Figure 7.3. The M:N

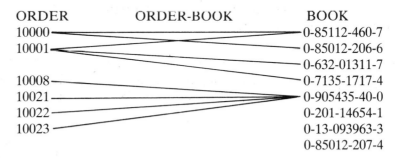

Figure 7.5(a)   Entity-relationship Occurrence Diagram Showing the Original M:N Relationship

Figure 7.5(b)   Replacement of the ORDER-BOOK Relationship by a New ORDER-LINE Entity

Order-Book relationship of Figure 7.3 has become an Order-Line entity connected to the Order and Book entities by the Has and For relationships, respectively.

Many-to-many relationships are problematic for at least two important reasons.

— Firstly, they may mask very unsatisfactory parts of an entity-relationship model. This is illustrated in Figure 7.7.

— The second reason is practical. Many data management systems do not support many-to-many relationships directly. Thus these relationships must be decomposed into two 1:N relationships before they can be mapped on to the software.

Decomposition of many-to-many relationships will not always be straightforward. Consider the example illustrated by the entity-relationship diagram in Figure 7.7.

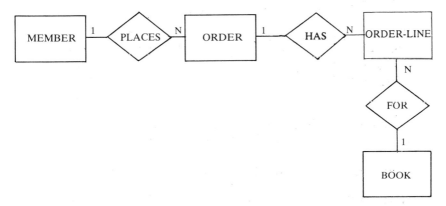

Figure 7.6   The New Entity-relationship Model Showing the M:N Decomposition

Figure 7.7   A Hidden M:N Relationship Restriction

The M:N relationship named Can Operate could imply either of two situations shown in Figures 7.8(a) and 7.8(b).

The original M:N relationship is subject to a hidden restriction. If a tutor can operate all those machines, but only those machines, where the tutor's specialism matches the specialism required by the machine, and vice versa, then the former decomposition (Figure 7.8(a)) is correct. However, if the relationship between Tutor and Machine is not constrained by Specialism in any way the latter decomposition (Figure

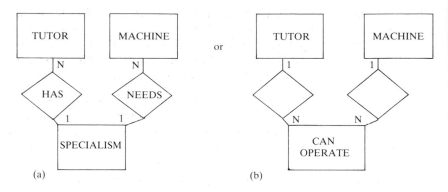

Figure 7.8   Alternative Decompositions of M:N Relationship

7.8(b)) applies. This is an important, but subtle, difference.

Finally, there is one further degree of relationship to consider, namely 1:1 (one-to-one). Use of this type of relationship is more likely to be used in the later stages of design, for example where some attributes of an entity are more often referenced than others. The Analyst may discover that some of the attributes of Member (such as member-name and member-address) are more frequently used than others (such as total-value-year-to-date, date-of-last-purchase and other management information data items). These latter data elements might be split off into a separate new entity. A new entity Member-Detail could be related to Member by a 1:1 relationship. This example would find expression in the following rule:

A member may have, at most, one member-detail.

A member-detail may belong to, at most, one member.

Figure 7.9 illustrates this rule.

Figure 7.9   An Entity-relationship Diagram Showing a 1:1 Relationship

7.2.5 Membership Class

A further property of the entity-relationship diagram which repays investigation is that of *membership class*. This is concerned with how the nature of the relationship is affected by obligatory and non-obligatory rules found in the problem under investigation. This should become clear by examining another relationship that is likely to occur in the Mail Order Book Club system, that between Book and Publisher.

It is possible to extend the Analyst's knowledge of the system by considering whether every occurrence of an entity must participate in a relationship. Howe (Howe, 1983) includes this additional information on the data model by using a blob or dot inside the entity symbol to signify that the entity's membership class is obligatory, whereas a blob or dot outside the entity symbol means that it is not obligatory.

Figure 7.10 Obligatory Membership

The diagram in Figure 7.10 expresses the rule:

A book must have at least one publisher.

A publisher must publish at least one stocked book.

Thus a book without a publisher is not permitted (how does the Analyst cope with internally produced reports?) and a publisher for whom the club has no books is also disallowed. The occurrence diagram in Figure 7.11 illustrates this type of membership class.

BOOK	BOOK-PUB	PUBLISHER
0-85112-460-7		Guinness Books
0-85012-206-6		NCC Publications
0-632-01311-7		Blackwell Scientific
0-7135-1717-4		G Bell & Sons Ltd
0-905435-40-0		DP Publications
0-201-14654-1		Addison-Wesley
0-85012-207-4		

Figure 7.11   Occurrence Diagram Illustrating Obligatory Membership

Figure 7.12   Non-obligatory Membership

The entity-relationship model in Figure 7.12 represents the rule:

A book need not have a publisher.

A publisher need not have published a stocked book.

These rules permit the Analyst to include internally produced reports in the model and also to record publisher details even where no books are currently stocked from that publisher. An occurrence example of this is shown in Figure 7.13.

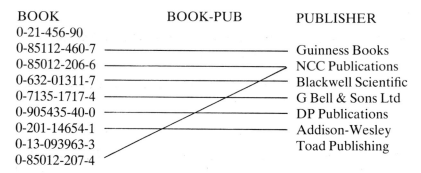

Figure 7.13   Occurrence Diagram Illustrating Non-obligatory Membership

Thus Book number 0-21-456-90 does not take part in the relationship Book-Pub because it is internally produced. Similarly there are no books stocked from Toad Publishing.

Figure 7.14   Non-obligatory/Obligatory Relationship

Figure 7.14 illustrates the rules:

A book need not have a publisher.

A publisher must publish at least one of the stocked books.

This can be further illustrated by the occurrence diagram in Figure 7.15.

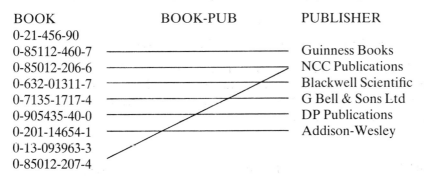

BOOK	BOOK-PUB	PUBLISHER
0-21-456-90		
0-85112-460-7		Guinness Books
0-85012-206-6		NCC Publications
0-632-01311-7		Blackwell Scientific
0-7135-1717-4		G Bell & Sons Ltd
0-905435-40-0		DP Publications
0-201-14654-1		Addison-Wesley
0-13-093963-3		
0-85012-207-4		

Figure 7.15   Occurence Diagram Illustrating a Non-obligatory/Obligatory Relationship

In this instance the publisher (Toad Publishing) is not permitted to take part in the model of the system.

Figure 7.16   Obligatory/Non-obligatory Relationship

Figure 7.16 covers the following circumstances:

A book must have at least one publisher.

A publisher need not have published a stocked book.

In this model (Figure 7.17) the internally produced report, given the number 0-21-456-90, is not permitted because it does not have a publisher. However, Toad Publishing can now appear in the model. Knowledge of membership classes will come in useful in the design of the file structures, as discussed in the companion text *Introducing Systems Design.*

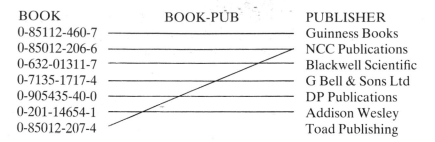

**Figure 7.17   Occurrence Diagram Illustrating an Obligatory/
Non-obligatory Relationship**

7.2.6   Entity-relationship Modelling: Summary

The basic concepts of entity-relationship modelling have been introduced. The diagrams highlight the entities that are important to the organisation (in the context of the study) and summarise the relationships that exist between those entities. The structure of the diagram is a direct reflection of the enterprise rules that govern the operation of the company. The organisational context of the model must never be forgotten or neglected.

The entity-relationship diagram helps the Analyst part way towards a data model, but it concentrates upon entities rather than attributes. The Analyst now needs to assign all the attributes discovered in the logical data flow model to the entities of the entity-relationship diagram. This assignment may lead to the discovery of new entities and also new attributes which have to be integrated into both logical views of the data. This effectively combines the strengths of the two ways of looking at data. The processes of the data flow diagram have produced an underlying data structure (in the data stores) that serves the processes, but may do so inflexibly and inefficiently. The structures provided by the entity-relationship model should be more stable and concise (because they are not purely driven by processes) but, at the same time, they should support those processes. The allocation of attributes to entities is well illustrated by Howe (Howe, 1983) who uses transactions (the processes of our data flow model) to test out the completeness of the entity-relationship diagram.

7.3   TABLES

The attributes of an entity may be simply represented in a table (sometimes called a relation). An example of such a table (called Member) from the Mail Order Book Club system is given in Figure 7.18. The table name is taken from the name of the entity.

Member-number	Member-name	Member-address
365487	Slack B	24 Albert Road
234567	Jones L	4 High Street
345677	Wolf J	1 The Rise
345623	Mills G	5 New Walk

Figure 7.18   The Member Table

In such a table each column name represents an attribute, with specific values of each attribute being shown in the rows. For example, Slack B is a value of the attribute member-name.

Although we have stressed the top-down development of tables it must be recognised that tables may alternatively be developed in a bottom-up manner from the attributes of the data stores and flows of the data flow diagram. Such tables may be complex and contain some redundant attributes whose removal will not result in any loss of information.

When constructing tables the Analyst should observe and apply a number of rules and restrictions:

1   The ordering of rows is not significant. No information is lost if, for example, the first row becomes the last.

2   The order of the columns is not significant.

3   Each row/column intersection contains only one value.

4   Each row in a table must be distinct so that it can be uniquely identified by quoting one or more attribute values.

A table that satisfies the restrictions listed above is in *first normal form*. The Analyst now proceeds to examine the table rigorously, passing through further stages of normalisation, until the data contained in the table is represented in a table or tables that are at least in *third normal form.* This is done by establishing whether each table has certain undesirable characteristics which will cause problems when the tables are physically implemented. This is the main reason for requiring normalisation. If the entities and relationships are to become physical tables or files (see the companion text *Introducing Systems Design* for detailed discussion) then the Analyst must not inadvertently build in features which will cause problems and anomalies in implementation. Tables which are not fully normalised will have these features and so the process of normalisation is very important.

7.3.1   Determinants and Identifiers

The table Member given above obeys the enterprise rule that:

A member-number may only have one associated member-name.

In such circumstances the attribute member-number is a determinant of the attribute member-name. Thus the value of the member-name is determined by the value of member-number.

Determinants may be composite. For example, in the Order-Line table in Figure 7.19 order-number and ISBN together determine quantity-ordered, as neither attribute would be sufficient on its own to determine a single value of quantity-ordered. The Analyst must ensure, however, that the composite determinant does not include a superfluous attribute. For example, member-address in the table Member may be

Order-number	ISBN	Quantity-ordered
10000	0-85112-460-7	12
10000	0-85012-206-6	2
10001	0-85112-460-7	1
10001	0-632-01311-7	21
10001	0-7135-1717-4	2
10008	0-905435-40-0	1
10021	0-905435-40-0	1
10022	0-905435-40-0	5
10023	0-905435-40-0	6

Figure 7.19   The Order-Line Table

determined by the composite determinant member-number, member-name. But member-number is sufficient and so member-name becomes superfluous.

The rule that each row in a table must be distinct means that an individual row can always be identified by quoting the values of all its attributes. In most cases it will not be necessary to know all the attribute values, but a limited subset should permit unique identification. This subset is known as the identifier. In the table Member it could be member-number. It is unlikely to be member-name because the Club is likely to have duplicated names (for example, five Smith J's).

In general an identifier is an attribute or combination of attributes whose value (or values) is sufficient to identify a row. A composite identifier may not contain superfluous attributes and none of the attributes of the identifier can contain null values. This is because the identifier effectively acts as the row's label and it seems reasonable to insist that the value of this label should be known. In the example member-number was introduced to act as a label or identifier. A composite identifier member-name, member-address) is a possible alternative but it is unlikely to guarantee uniqueness (Joe Smith and his son John live at the same address) and it is unnecessarily clumsy. Moreover, it is liable to change (for example, a Member may move house).

7.3.2 Normalisation of Tables

Normalisation is a technique used by the Analyst to identify a collection of two-dimensional tables which are independent and contain no unnecessary redundant data. The tables resulting from normalisation process contain simple data items with relationships being represented by replication of data items. Normalisation ensures that insertions, deletions and amendments may then be made to the data without undesirable consequences. There may also be a saving on data storage requirements as a result of using normalised tables.

The problems of tables which are not fully normalised are illustrated using a new table (called Order-UN in Figure 7.20) from the data store for pending orders. This will be examined and taken through the stages of normalisation.

The top two lines of Figure 7.20 are logically a single row in the table as they have the same identifier (an Order number of 10000). Similarly,

Order-number	Member-number	Member-name	Member-address	Order-date	ISBN	Quantity-ordered	Title	Price £
10000	234567	Jones S	4 High St	11/12/87	0-85112-460-7	12	MIS Concepts	19.95
					0-85012-206-6	2	Systems Analysis	8.50
10001	345623	Mills G	5 New Walk	12/12/87	0-85112-460-7	1	MIS Concepts	19.95
					0-632-01311-7	21	Data Analysis	6.99
					0-7135-1717-4	2	Audit Controls	7.00
10022	345677	Wolf J	1 The Rise	21/12/87	0-905435-40-0	5	Costing	14.50
10023	345677	Wolf J	1 The Rise	22/12/87	0-905435-40-0	6	Costing	14.50

Figure 7.20   Order-UN: the Unnormalised Table

lines 3–5 constitute a single logical row. Consequently, there are multiple values of each of the first two logical rows where they intersect the ISBN, Quantity-ordered, Title and Price columns.

The first stage of normalisation (conversion to first normal form) is to ensure that all the row/column intersections contain only one value by removing repeating groups to form separate tables. In the example, Order-UN is said to be *unnormalised* because there are several values for the group of attributes ISBN, quantity-ordered, title and price associated with the row identifier order-number. These should be separated into another table (Figure 7.21). The identifier of the new table (called Order-Line-1NF) is the combination of order-number and ISBN. The remaining attributes from Order-UN constitute another table, called Order-1NF (Figure 7.22). Note that the order-number attribute appears in both the new tables, so linking them together.

Order-number	ISBN	Quantity-ordered	Title	Price £
10000	0-85112-460-7	12	MIS Concepts	19.95
10000	0-85012-206-6	2	Systems Analysis	8.50
10001	0-85112-460-7	1	MIS Concepts	19.95
10001	0-632-01311-7	21	Data Analysis	6.99
10001	0-7135-1717-4	2	Audit Controls	7.00
10022	0-905435-40-0	5	Costing	14.50
10023	0-905435-40-0	6	Costing	14.50

Figure 7.21   Order-line-1NF, a First Normal Form Table

Order-number	Member-number	Member-name	Member-address	Order-date
10000	234567	Jones S	4 High St	11/12/87
10001	345623	Mills G	5 New Walk	12/12/87
10022	345677	Wolf J	1 The Rise	21/12/87
10023	345677	Wolf J	1 The Rise	22/12/87

Figure 7.22   Order-1NF

Although the complexity of the repeating group has been eliminated, Order-1NF and Order-Line-1NF still pose certain problems. For example, Order-Line-1NF suffers from the following problems when updating is required:

Insert   No attribute in the identifier may have a null value, so this means that both order-number and ISBN must have entries in the Order-Line-1NF example given. Thus a book cannot appear in the table until it has been ordered. Hence the title and price of book ISBN 0-201-14654-1 is not known as this book has not been ordered.

Delete   If an order for a certain book is completed then that order is no longer pending and must be deleted. This means that the whole row must be deleted to avoid contravening the rule that the table cannot have a null value in the identifier. If a particular row is deleted information is lost about the specific order and about the book. For example, if the fifth row of the table Order-Line-1NF is deleted the title and price of book ISBN 0-7135-1717-4 is lost.

Amend   If the price of a book requires changing it is necessary to search for all occurrences referring to the book. Missing one would lead to inconsistency in the table. For example, the first and third rows of Order-Line-1NF could contain two different prices for book ISBN 0-85112-460-7.

The table Order-Line-1NF may be improved by further splitting. The general guideline behind this splitting is to remove to another table attributes which are both not in the identifier and not determined by the whole of the identifier. Thus, in this example, attributes title and price are not in the identifier and are also not determined by the whole of the identifier because order-number is irrelevant. Thus they may be split away. This again creates two tables. These new tables are in second normal form. In fact, as explained later, in this example these two tables are also already in third normal form, so they have been named Order-Line-3NF and Book-3NF.

This has permitted the insertion of the book ISBN 0-201-14654-1 which is currently not on order. It will also permit the title and price of, for example book ISBN 0-7135-1717-4 to be retained after order-number 10001 has been deleted. Updating is also made easier. If the price changes then only one row of the table Book-3NF has to be updated. The cost of these improvements is the repetition of the column order-number, now held in both Order-1NF and Order-Line-3NF.

ISBN	Title	Price £
0-85112-460-7	MIS Concepts	19.95
0-85012-206-6	Systems Analysis	8.50
0-632-01311-7	Data Analysis	6.99
0-7135-1717-4	Audit Controls	7.00
0-905435-40-0	Costing	14.50
0-201-14654-1	Data Protection	3.50

Figure 7.23   Book-3NF, a Third Normal Form Table

Order-number	ISBN	Quantity ordered
10000	0-85112-460-7	12
10000	0-85012-206-6	2
10001	0-85112-460-7	1
10001	0-632-01311-7	21
10001	0-7135-1717-4	2
10022	0-905435-40-0	5
10023	0-905435-40-0	6

Figure 7.24   Order-line-3NF

What the Analyst discovers in this procedure is a flaw in his original entity-relationship modelling. The earlier model did not recognise Order-Line and Book as two separate entities, but this error has been picked up by the second stage of normalisation so the model may now be corrected. The tables Book-3NF and Order-Line-3NF may be left because they are already in a further stage of normalisation – *third normal form*. However the table Order-1NF still has some undesirable features. These are revealed when tests of *insert, delete* and *amend* are applied. This table is reproduced in Figure 7.25. The identifier of Order-1NF is order-number.

Order-number	Member-number	Member-name	Member-address	Order-date
10000	234567	Jones S	4 High St	11/12/87
10001	345623	Mills G	5 New Walk	12/12/87
10022	345677	Wolf J	1 The Rise	21/12/87
10023	345677	Wolf J	1 The Rise	22/12/87

Figure 7.25   Order-1NF

Insert   The table cannot show a particular member has a certain member-address until an order has been received.

Delete   If an order is deleted the information about the member's name and address is lost. This effectively destroys the concept of Book Club membership. Moreover, a banned member may have his number cancelled but name and address still has to be available to help recover previous debts.

Amend   If a member changes his address (or name) then it is necessary to search for all orders for this member. Missing one would lead to inconsistency in the table.

Third normal form is concerned with producing tables which are (a) in second normal form, and (b) have no dependencies between the attributes which are not in the identifier. In this instance member-number, which is not part of the identifier, determines member-name and member-address, so these can be split off from the table Order-1NF. Thus Order-1NF is split again into two tables, Order-3NF and Member-3NF.

Order-number	Member-number	Order-date
10000	234567	11/12/87
10001	345623	12/12/87
10022	345677	21/12/87
10023	345677	22/12/87

Figure 7.26   Order-3NF

Member-number	Member-name	Member-address
234567	Jones S	4 High St
345623	Mills G	5 New Walk
345677	Wolf J	1 The Rise

Figure 7.27   Member-3NF

The attributes of the final third normal form tables (those with -3NF

suffix to the name) are summarised below. The attributes underlined are
the identifiers of the tables.

ORDER :   (order-number, member-number, order-date)

BOOK :   (ISBN, title, price)

ORDER-LINE :   (order-number, ISBN, quantity-ordered)

MEMBER: (member-number, member-name,
                 member-address)

By examining the way in which the attributes order-number,
member-number and ISBN link the tables together, it is possible to
derive the following entity-relationship diagram (Figure 7.28). This

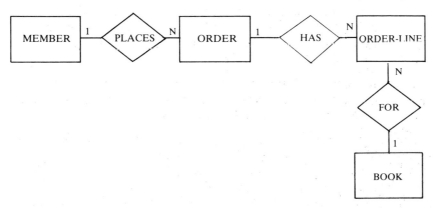

Figure 7.28   Mail Order Book Club Entity-relationship Diagram

diagram is identical to that previously derived by a more intuitive
top-down approach (Figure 7.6).

The progression from first to third normal form has been given so
that the problems of inadequately normalised tables are fully appreciated. However, it is possible to create normalised tables directly by
applying the principle known as Boyce/Codd Normal Form (BCNF).
This is actually slightly more powerful than third normal form and
merely requires that in a table "every determinant must be a candidate
identifier". In other words, an attribute that is a determinant of another
attribute must be a possible identifier. If it is not, then new tables should
be created where the non-identifying determinant becomes a possible
identifier in a new table.

For example, in the table Order-1NF the attribute member-number determined member-name and member-address but it is not a possible identifier because its values do not permit the unique identification of a row. Member-number 345677 occurs in two rows. Hence the table is not in BCNF. This can be remedied by splitting the table and constructing a new table where this determinant is a possible identifier.

Mastering this concept provides the Analyst with a quicker way of producing normalised tables than ploughing through separate stages of normalisation. It is also worth knowing that fourth and fifth normal forms exist; but this is a matter for discussion in a specialised text. Interested readers should consult the references to this chapter.

7.3.3 Normalisation: Summary and Notation

Normalisation is concerned with producing a robust table structure that will support the implementation of the data model. It ensures that the Analyst does not build unnecessary and damaging side effects into a system. Because it is built independently of any reference to software it may successfully provide the basis for mapping on to any suitable package or language. In simple terms, the tables become the candidate file structures where each attribute is a field and each row a record. Within each file there will be an identifier (candidate key field(s)) which can be used to uniquely identify a particular record. It should also be clear that no information is lost by splitting the tables. Information available from the table Order-UN can still be found by accessing tables Member-3NF, Order-3NF, Order-Line-3NF and Book-3NF. If the Analyst loses information along the way then the table split has been incorrect.

7.4 DATA ANALYSIS AT INFOSYS

7.4.1 Mail Order Book Club

The entity-relationship model shown in Figure 7.29 was developed after undertaking a detailed analysis of the requirements of the Mail Order Book Club (introduced in Chapter 6). Attributes from the data stores and data flows of the logical model (Figure 6.9) need to be assigned to the top-down entity-relationship model. The tables derived from this model will need to support at least the transactions listed below. Others will be found in both the Level 0 and lower-level data flow diagrams. However, a limited list has been selected for illustrative purposes (Figure 7.30).

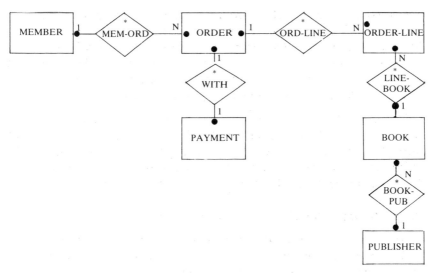

Figure 7.29   Mail Order Book Club Entity-relationship Diagram

* Verification and clearance of order details
* Order consolidation and production of picking instructions
* Generation of member order despatch details
* Compilation of the offer list
* Compilation of a member mailshot and labels.

Figure 7.30   Order Processing Transaction List

The managers of InfoSys also have several queries they would like the computer system to be able to respond to. For instance, which books have been the most popular this month?

The enterprise rules for this system are:

A member makes one or many orders for one or more copies of one or more books! Each order should be accompanied by one payment due. Each book has only one publisher.

The initial entity tables are suggested in outline only. The dots illustrate unallocated attributes.

MEMBER :   (member-number, member-name,
                member-address, join-date, ...)

ORDER :   (<u>order-number,</u> order-date, despatch-date, ...)

ORDER-LINE : (<u>order-number,</u> <u>ISBN,</u> quantity-ordered, ...)

BOOK :   (<u>ISBN,</u> title, author, selling-price,
cost-price, quantity-in-stock, re-stock-level, ...)

PAYMENT :   (<u>transaction-number,</u> payment-amount,
payment-method, clearance-date . . . .)

PUBLISHER : (<u>publisher-name,</u> publisher-address . . . .)

All the relationships (except With) are of the 1:many type. These relationships can be represented by posting the identifier of the entity table at the '1' side into the entity table at the 'many' side. Thus the identifier member-number is posted into the Order table to establish the link, removing the need for a separate relationship table. Similarly publisher-name is posted into the Book table. As order-number is already part of the composite identifier of the Order-Line entity table, posting it again would be superfluous. Similarly, ISBN is already posted (ie it is preposted) from the Book table into the Order-Line table. The 1:1 relationship With can be represented either by posting order-number into the Payment table, or by posting transaction-number into the Order table. Yet another solution would be to combine the Order and Payment tables into a single table. The choice of solution will be governed by the nature of the processing required.

The relationships which do not require separate tables because identifiers have been posted are indicated on the entity-relationship diagram by an asterisk.

The full set of normalised tables for the Mail Order Book Club is now as follows. The posted and preposted identifiers are printed in bold.

MEMBER :   (<u>member-number,</u> member-name,
member-address, join-date)

ORDER :   (<u>order-number,</u> order-date, despatch-date,
member-number)

ORDER-LINE : (**<u>order-number, ISBN,</u>** quantity-ordered)

BOOK :   (<u>ISBN,</u> title, author, selling-price, cost-price,
quantity-in-stock, re-stock-level, **publisher-name**)

PAYMENT :   (<u>transction-number,</u> **order-number,**

payment-amount, payment-method,
clearance-date)

PUBLISHER :   (<u>publisher-name</u>, publisher-address)

7.4.2   Seminar Bookings

The seminar bookings system also requires detailed examination to
identify further relevant entities and associated attributes. In this case
the entity-relationship diagram and tables which result from the data
analysis need to be able to support at least the requirements shown in
Figure 7.31.

* Acceptance of seminar bookings
* Generation of a delegate list for a given seminar
* Generation of a schedule of future seminars
* A list of delegates requiring accommodation for a given seminar
* Production of a confirmation of booking to be sent to each delegate

Figure 7.31   Seminar Bookings Transaction List

It is also desirable that the data model be able to support queries such
as those shown in Figure 7.32.

* How many delegates have been booked on a particular seminar?
* What alternative dates can be offered for a fully booked seminar?

Figure 7.32   Additional Transactions

The seminar booking schedule has been expressed as a table called
Seminar-booking-UN (Figure 7.33) which results from one of the data
flows of the system. It will be used to illustrate both the development of
the entity-relationship diagram and the normalisation process. The data
analysis will be illustrated from both a top-down and a bottom-up
perspective.

Booking-number	Booking-date	Delegate-number	Delegate-name	Delegate-address	Seminar-number	Title	Date	Venue
5487	13/05/87	58	Martinek M	14 High St Oxford	1254	Systems Analysis	18/06/87	Manchester
					1255	Systems Design	21/06/87	London
5488	14/05/87	87	Jones L	4 Moor Lane Reading	1254	Systems Analysis	18/06/87	Manchester
5489	14/05/87	95	Morris C	48 Queen St Leeds	1260	Data Analysis	19/06/87	Reading
5490	14/05/87	95	Morris C	48 Queen St Leeds	1255	Systems Design	21/06/87	London
5491	15/05/87	24	Brown P	3 Dumps Rd Cheam	1254	Systems Analysis	19/06/87	London

Figure 7.33   Seminar-booking-UN

Top-down Analysis

It seems likely that Seminar, Delegate and Booking are highly probable candidates for entities. Information is required about delegates who make bookings for the seminars they wish to attend. The first stage of entity modelling is given in Figure 7.34.

Consider first the Delegate and Booking entities. A delegate cannot be uniquely identified by his or her name and so a new identifier, delegate-number, must be used. A further obvious attribute associated with delegate is delegate-address which is needed to send registration details. Information is also required about bookings. For instance the date a booking was made and whether the delegate wishes to be residential or non-residential. Bookings are related to delegates, as a particular booking must be for a particular delegate.

The Analyst now needs to know the degree of the relationship between these two entities. This will require an understanding of how the system works, that is the enterprise rules. It is possible that a delegate may make separate bookings for several seminars which he or she wishes to attend. This is recorded in the Seminar-Booking table by different line entries. For instance the delegate M Martinek made a single booking for two different scheduled seminars. However, one booking can only be made by one delegate, and so the relationship between booking and delegate is 1:N.

A booking is also related to a seminar, since information is required about which seminar the booking was made for. The Seminar entity will

Figure 7.34   Initial Seminar Booking Entity-relationship Diagram

be indentified by the seminar-number and its date. The relationship between Booking and Seminar is M:N. for example M Martinek made a booking on 13/05/87 for two different seminars (Systems Analysis and Systems Design) which are planned to take place on different dates. A particular seminar may have many bookings made for it. For example the Systems Analysis seminar held on 18/06/87 has bookings 5487 and 5488.

The remaining relationship degrees can now be inserted in the entity-relationship diagram (Figure 7.35).

Figure 7.35   Seminar Booking Entity-relationship Diagram

At first this model seems to work well in representing the structure of the data. However, a seminar can be scheduled to take place on several occasions and it is an intention of the seminar manager to be able to enquire about the most popular seminars. The model shown in the diagram (Figure 7.36) was developed on closer scrutiny of the data. A booking may still be made for many seminars but these are actually scheduled seminars on particular dates. Each scheduled seminar (identified by seminar-number and date) is of a particular seminar type (identified by seminar-number) and there may be many scheduled seminars for any one seminar type. Consequently, there is a 1:N relationship between Seminar Type and Scheduled Seminar.

A further area of investigation concerns the tutors who will be responsible for the various scheduled seminars. A list of the tutors who have been assigned to the scheduled seminars is required by the seminar manager for planning the future events. The Tutor entity is connected to the Scheduled Seminar entity by a 1:many relationship since a scheduled seminar is the responsibility of one tutor, but since a scheduled seminar lasts only a few days a tutor can be allocated to many

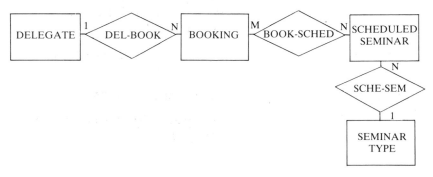

Figure 7.36   Revised Seminar Bookings Entity-relationship Diagram

scheduled seminars. Figure 7.37 includes the extension to cover tutors and also shows the membership classes.

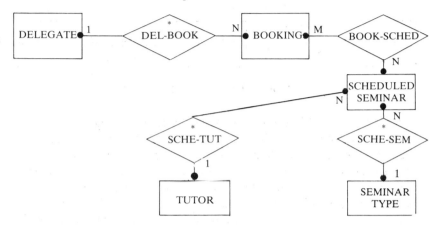

Figure 7.37   Detailed Seminar Bookings Entity-relationship Diagram

The tables derived from the entities in Figure 7.37 are outlined below in Figure 7.38. Some other attributes have been added to reflect, amongst other things, posted and pre-posted identifiers (shown in bold print). Further analysis would be required to complete the tables.

SEMINAR TYPE :   (<u>seminar-number, title</u>, ...)

SCHEDULED SEMINAR :   **(<u>seminar-number, date,</u> venue,
                          tutor-name</u> . . . .)**

BOOKING :   (<u>booking-number</u>, amount-due,
                          amount-paid, booking-date,
                          delegate-number . . . .)

DELEGATE :   (<u>delegate-number</u>, delegate-name,
                          delegate-address . . . .)

TUTOR :   (<u>tutor-name</u>, tutor-grade . . . .)

BOOK-SCHED :   **(<u>booking-number, seminar-number,
                          date</u> . . . .)**

Figure 7.38   Outline Seminar Booking Tables

In the case of the many-to-many relationship Book-Sched, the relationship table will consist of the identifying attributes of the entities Booking and Scheduled Seminar. All the other relationships (marked with asterisks in Figure 7.37) are of the 1:many type and are represented by posting the identifiers of the entity table at the 1 side into the entity table at the many side. Thus the identifier delegate-number is posted into the Booking table to establish the link, removing the need for a separate relationship table, Dcl-Book. Similarly, tutor-name is posted into the Scheduled Seminar table. Seminar-number is already part of the composite identifier of Scheduled Seminar and thus posting it again would be superfluous.

Bottom-up Analysis

In contrast to the above top-down analysis, we now consider a purely bottom-up approach. It is assumed (for pedagogic reasons only) that the Analyst initially starts with a single entity Booking, and places all the attributes from the seminar booking table into it. The table, Booking-UN (Figure 7.39) which is the same as Seminar-Booking-UN in Figure 7.33, is found on examination to be un-normalised since a value of the identifier booking-number may be associated with a repeating group of seminar attributes. The process of normalisation will split up this large table into several concise tables which do not duplicate attribute values unnecessarily.

The attributes seminar-number, title, date and venue are a repeating group in the table Booking-UN. These attributes should be split into a separate table called Booking-Seminar-1NF which also contains the attribute booking-number to maintain the link between the date in each table. The identifier of Booking-Seminar-1NF (Figure 7.40) is now booking-number, seminar-number and date. It is in first normal form as the identifier now determines a single value for each of the table attributes.

The remaining table after the split is now called Booking-2NF (Figure 7.41). It is already in second normal form as there are no repeating groups and because the identifier is a single attribute (booking-number) there can be no dependencies on part of the key.

The table Booking-Seminar-1NF (Figure 7.40) can be improved by further splitting. The value of the attribute Title is not dependent on the full identifier. Title is determined by the value of seminar-number only,

Booking-number	Booking-date	Delegate-number	Delegate-name	Delegate-address	Seminar-number	Title	Date	Venue
5487	13/05/87	58	Martinek M	14 High St Oxford	1254	Systems Analysis	18/06/87	Manchester
					1255	Systems Design	21/06/87	London
5488	14/05/87	87	Jones L	4 Moor Lane Reading	1254	Systems Analysis	18/06/87	Manchester
5489	14/05/87	95	Morris C	48 Queen St Leeds	1260	Data Analysis	19/06/87	Reading
5490	14/05/87	95	Morris C	48 Queen St Leeds	1255	Systems Design	21/06/87	London
5491	15/05/87	24	Brown P	3 Dumps Rd Cheam	1254	Systems Analysis	19/06/87	London

Figure 7.39   Booking-UN

Booking-number	Seminar-number	Title	Date	Venue
5487	1254	Systems Analysis	18/06/87	Manchester
5487	1255	Systems Design	21/06/87	London
5488	1254	Systems Analysis	18/06/87	Manchester
5489	1260	Data Analysis	19/06/87	Reading
5490	1255	Systems Design	21/06/87	London
5491	1254	Systems Analysis	19/06/87	London

Figure 7.40 Booking-Seminar-1NF

not booking-number, seminar-number and date. The attributes seminar-number and title should therefore be split into a further table called Seminar-Type-3NF (Figure 7.42) where the identifier is only seminar-number.

Booking-number	Booking-date	Delegate-number	Delegate-name	Delegate-address
5487	13/05/87	58	Martinek M	14 High St Oxford
5488	14/05/87	87	Jones L	4 Moor Lane Reading
5489	14/05/87	95	Morris C	48 Queen St Leeds
5490	15/05/87	95	Morris C	48 Queen St Leeds
5491	15/05/87	24	Brown P	3 Dumps Rd Cheam

Figure 7.41 Booking-2NF

The splitting of Booking-Seminar-1NF leaves a table called Booking-Seminar-1NF* (Figure 7.43) with the identifier booking-number, seminar-number, date. However this is not yet in second normal form as there is still an attribute, venue which is dependent upon part of the identifier, in this case seminar-number and date. The Analyst would split the table Booking-Seminar-1NF still further into two tables which can now be identified as being in at least second normal form. These tables are Scheduled-Seminar-3NF (Figure 7.44) with the identifier seminar-number and date and Booked-Scheduled-Seminar-3NF (Figure 7.45) with identifier booking-number, seminar-number and date. Note that they are already in third normal form as they contain no dependencies between non-identifying attributes.

Seminar-number	Title
1254	Systems Analysis
1255	Systems Design
1260	Data Analysis

Figure 7.42   Seminar-Type-3NF

Booking-number	Seminar-number	Date	Venue
5487	1254	18/06/87	Manchester
5487	1255	21/06/87	London
5488	1254	18/06/87	Manchester
5489	1260	19/06/87	Reading
5490	1255	21/06/87	London
5491	1254	19/06/87	London

Figure 7.43 Booking-Seminar-1NF*

Seminar-number	Date	Venue
1254	18/06/87	Manchester
1255	21/06/87	London
1260	19/06/87	Reading
1254	19/06/87	London

Figure 7.44   Scheduled-Seminar-3NF

Returning to the earlier table Booking-2NF (Figure 7.41) to complete the normalisation process, this table is not yet in third normal form as there is a dependency between the attribute delegate-number which is not part of the table identifier and the attributes delegate-name and delegate-address. This group of attributes should therefore be split off into a new table Delegate-3NF (Figure 7.46) whose identifier is delegate-number. The remaining attributes form a new table called Booking-3NF (Figure 7.47) where the identifier is still booking-number.

Booking-number	Seminar-number	Date
5487	1254	18/06/87
5487	1255	21/06/87
5488	1254	18/06/87
5489	1260	19/06/87
5490	1255	21/06/87
5491	1254	19/06/87

Figure 7.45   Booked-Scheduled-Seminar-3NF

Delegate-number	Delegate-name	Delegate-address
58	Martinek M	14 High St Oxford
87	Jones L	4 Moor Lane Reading
95	Morris C	48 Queen St Leeds
24	Brown P	3 Dumps Rd Cheam

Figure 7.46   Delegate-3NF

Booking-number	Booking-date	Delegate-number
5487	13/05/87	58
5488	14/05/87	87
5489	14/05/87	95
5490	15/05/87	95
5491	15/05/87	24

Figure 7.47   Booking-3NF

The entity-relationship model, shown in Figure 7.36, corresponds to this collection of tables in Third Normal Form. A more comprehensive representation of the requirements of the system would clearly need to include entities such as Tutor and Seminar-Texts.

The tables for the seminar booking data model are summarised in

Figure 7.48. Comparison of these tables with the entity-relationship diagram in Figure 7.36 and the tables in Figure 7.38 shows that the top-down and bottom-up analyses have produced essentially the same result.

SEMINAR-TYPE-3NF :	(<u>seminar-number</u>, title)
SCHEDULED-SEMINAR-3NF :	(<u>seminar-number</u>, <u>date</u>, venue)
BOOKING-3NF :	(<u>booking-number</u>, booking-date, delegate-number)
DELEGATE-3NF :	(<u>delegate-number</u>, delegate-name, delegate-address)
BOOKED-SCHEDULED-SEMINAR-3NF :	(<u>booking-number</u>, <u>seminar-number</u>, <u>date</u>)

Figure 7.48   Third Normal Form Seminar Booking Tables

7.5   DATA ANALYSIS: SUMMARY

This chapter has introduced the techniques of Data Analysis. The following framework is suggested for applying these techniques to the construction of a logical data model.

1   Begin by writing down a preliminary list of those entity types which you can confidently identify in the problem area under considera-tion. Select an identifier for each of these and write this down in a skeleton table which just shows the entity type and the identifier. Do not try to be over-ambitious at this stage.

2   Sketch a simple entity-relationship model that shows the known relationships between the entity types. Include relationship degree and membership class details.

3   Use a data flow diagram and its supporting data dictionary to prepare a list of processes that should be supported, and the data elements or items that must be recorded.

4   Make a preliminary allocation of attributes to entities. Check to see if all the processes required by the data flow diagram can be supported. Extend the entity-relationship model as necessary to cope with processes that are not supported and attributes that cannot be allocated to present tables.

5   Check that the entities, relationships and attributes are all appropriate. Check that the set of tables are fully normalised and confirm that all the processes are supported by both the entity-relationship diagram and the table structure.

A more rigorous method for building the model is presented in Howe (Howe, 1983), whilst Date's book provides more detail on the process and pitfalls of normalisation (Date, 1986).

However, what should be clear from the approach suggested above is that there is a need for consistency and iteration. The entity-relationship model has to be checked to see if it supports all the processes and data elements of the data flow model, and has to be altered until it does. This procedure is an ideal candidate for automation. Furthermore, if the enterprise rules can be succinctly and accurately summarised then normalisation can be performed automatically. Automation aids the production and editing of the diagrams, resulting in the fast development of verifiable models. Such tools are already provided in some Analyst Workbenches. Furthermore, a major Italian project (Albano *et al*, 1985) sets out a methodology for data design specifically supported by a set of automated tools. An annotated bibliography on automated aids in this area lists about forty papers and texts (Holley, 1987). It is undoubtedly an area of great promise and activity.

This chapter has:

— Outlined the need for a technique to model the system from a static logical perspective.

— Described the modelling conventions of entity-relationship diagrams and discussed their construction.

— Explained the purpose and procedures of normalisation.

— Illustrated entity-relationship modelling and normalisation within the context of the case study.

— Contrasted the top-down and bottom-up development of an entity-relationship model.

7.6   PRACTICAL EXERCISES AND DISCUSSION POINTS

1   Draw an entity-relationship diagram and skeleton tables for the City Dyers case study introduced in Chapter 5.
    Extract the necessary attributes from the documents used and the

output requirements of management and specify the normalised tables required to support a system which will:

— monitor work in progress;

— maintain customer shades for fabrics;

— generate invoices for completed orders.

2 Normalise the following table.

Examination invigilation schedule

Room	Date	Time	Course No	Title	Examination	Students	Invigilator	Staff Room
W5.13	11/06/87	14.00	125	Comp Sci	Programming	80	Skidmore	W2.17
W6.11	12/06/87	14.00	125	Comp Sci	Systems	80	Mills	W8.24
B1.23	12/06/87	09.00	207	Bus Stud	Modelling	25	Smith	W5.7
W5.13	12/06/87	14.00	402	Maths	Lin Algebra	32	Henton	B6.14
W6.11	13/06/87	14.00	207	Bus Stud	Systems	25	Mills	W8.24
W5.13	13/06/87	14.00	125	Comp Sci	Networks	80	Henton	B6.14
B1.23	14/06/87	09.00	207	Bus Stud	Programming	25	Skidmore	W2.17

3 A farmer wishes to keep computerised records on the milk and calf production of his dairy herd. All calves produced are sold and not added to the dairy herd. Each cow has a name and date of birth and will produce milk for a lactation period after the birth of a calf or calves. Milk recordings for each cow in terms of litres are taken each day. The information required for each pregnancy of a cow are the bull's name, date of mating, date of birth and each calf's weight at birth and sex. The system is to provide the following information to the farmer:

(a) The details of all births of calves attributed to each bull.

(b) The milk yield of a cow over a particular lactation period.

Draw an entity-relationship diagram for the system and suggest the tables which will support the farmer's requirements for information.

8 Supporting Tools: Data Dictionaries and Entity Life Histories

8.1 INTRODUCTION

The last two chapters have introduced models which have had significant effect on the teaching and practice of systems analysis and design. This chapter briefly examines two further techniques which support the understanding of the logical information system requirements of an organisation.

The first is the data dictionary. This holds the corporate data resource and provides a detailed underpinning of the other models introduced in this text. Chapter 6 has already shown how the data flow diagram must be supported by the detail of the data dictionary. It also plays an important role in the design of a physical implementation of the logical model.

Secondly, the entity life history is a method of pulling together the process driven data flow diagram and the static data model of data analysis. In doing so it permits the verification of these views within the modelling of time and sequence.

8.2 DATA DICTIONARY

The concept and contents of a data dictionary were introduced in Chapter 6 and examples were given for the data flow model. It was suggested that these might be maintained in a manual system but that considerable advantages were likely to accrue from using an automated Data Dictionary System (DDS). This DDS software (usually referred to as data dictionary in this text) can effectively and efficiently maintain a large central repository of data about the data of an organisation – so-called metadata. Metadata is a level of abstraction higher than the actual data used in operations; it is not the actual data used but data about the actual data used. Figure 8.1 shows the distinction between the two levels of data. Metadata is used to define, identify and describe the characteristics of the user data. Metadata usually falls into two categories.

— What the data is or what it means.

— Where the data can be found and how it can be accessed.

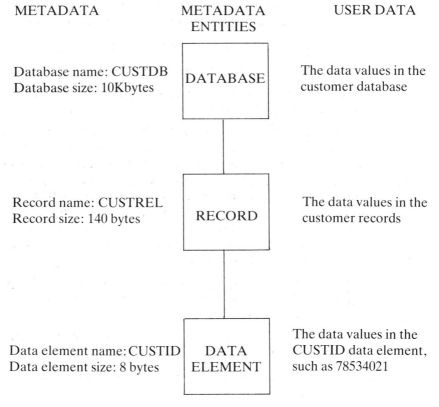

METADATA	METADATA ENTITIES	USER DATA
Database name: CUSTDB Database size: 10Kbytes	DATABASE	The data values in the customer database
Record name: CUSTREL Record size: 140 bytes	RECORD	The data values in the customer records
Data element name: CUSTID Data element size: 8 bytes	DATA ELEMENT	The data values in the CUSTID data element, such as 78534021

Figure 8.1   Metadata Versus Oser Data
(from Leong-Hong and Plagman, 1982)

There has recently been an increased acceptance of the benefits of establishing and maintaining a data dictionary and a number of products are available in the commercial market-place. A comparative review of 13 data dictionary systems was made by Mayne (Mayne, 1984). The central role of data dictionaries in commercial software and data processing organisation and standards demands that the concept be examined in greater depth.

8.2.1 Data as a Resource

Organisations are increasingly recognising that data is an important resource which, like other resources such as staff and materials, has both value and cost to the enterprise. Consequently data needs to be managed so that it effectively serves the information systems of the organisation. For this management to be successful there needs to be:

— Knowledge of what data exists and how it is used.

— Control of modifications to existing data or processes using data.

— Control over plans for new uses of data and over the acquisition of new types of data. (BCS, 1977.)

This need for control is increased where data is used for more than one application and the trend towards database solutions naturally means that this is often the case. In such circumstances the central definition of data and its use is usually delegated to a Data Administration function. The Data Administrator will be concerned with the correct use and maintenance of data, the integration of new applications and amendments, and the implementation details of data storage, access and manipulation. The data dictionary is an essential support tool for the successful performance of these tasks.

The data dictionary may be viewed as aiding most analysis and design activities. The BCS Working Party report (BCS, 1977) suggested that it could be used during each of the following stages:

— data analysis, to determine the fundamental structure of the data of the enterprise;

— functional analysis, to determine the way in which events and functions use data;

— database or conventional file design;

— transaction or program design;

— storage structure design, where this is a further refinement of the initial database or file design;

— operational running of application systems;

— collection and evaluation of performance statistics;

— database tuning to improve performance;

— application maintenance and database restructuring. (BCS, 1977.)

The following list of possible application areas given by Leong-Hong and Plagman also covers most activities in the systems development process:

— systems planning;

— requirements definition and analysis;

— design;

— implementation, programming, testing and conversion;

— documentation and standards;

— operational control and audit trail;

— end-user support. (Leong-Hong, 1982.)

The first type of data dictionary was a clerical card index system keeping track of details of the physical implementation. Information held on files and records used by programs made it easier to identify programs which needed recompiling in the event of a change in record or file structure (Redfearn, 1987). The advent of Data Base Management Systems (DBMS) with their emphasis on shared data reinforced this need for documentation. A data dictionary provided a means for ensuring standard terminology and effective documentation control through cross-referencing. The automation of this dictionary was a natural step.

The first automated data dictionary systems were primarily concerned with documenting the physical aspects of data processing – systems, programs, files and data bases. However, their scope has now increased to storing and manipulating logical models – such as data flow and entity-relationship diagrams. In doing so the data dictionary has progressed from a passive documentor of systems to an active productivity tool.

The activity of a data dictionary system demands scrutiny. All data processing applications require metadata to operate (such as database schemas, Cobol File Definitions and Job Control Languages) and an active data dictionary controls this processing environment. Indeed it is the scope of the metadata management that determines the activity of the data dictionary because the DDS is only active "with respect to a program or process if and only if that program or process is fully dependent upon the data dictionary for its metadata". (Plagman, 1978.) In a passive system metadata may be defined from other sources and the data dictionary acts as a documentation facility rather than an active tool in system development. In general, commercial DDS may be placed upon a spectrum of activity with the trend towards active or potentially active data dictionary systems.

This concept of activity is an example of how the term 'data dictionary' is used by vendors to describe software with a wide range of capabilities and facilities. Comparisons across products reveal marked differences in functionality. Consequently, the next section describes desirable features of a data dictionary system with the aim of providing an appreciation of the scope of DDS as well as a framework in which to assess competitive products. It is largely based upon the BCS Data Dictionary Systems Working Party report of 1977 which still remains an important benchmark in this field.

8.2.2 Desirable Features of a Data Dictionary

The BCS Working Party recommended that the data dictionary should operate at two distinct levels. The first is the logical level that gives the ability to record and analyse requirements irrespective of how they are going to be met. This logical view represents an implementation independent view of the enterprise and initial and successive implementations must take place within the scope of this framework. At least four benefits may be obtained from building this logical model:

— a perspective for system planning;

— an appreciation of how systems interact;

— a method of communicating complexity;

— a database design tool.

The second level of the data dictionary is the implementation level. This gives the facility to record physical design decisions in terms of the implemented database or file structures and the programs that access them. If the logical level is how the data is seen from the enterprise, then the implementation level is how the data is viewed by the file handling system or the DBMS itself.

Logical Level

The logical view describes the nature of the enterprise and its data. It is a model of the organisation showing things of interest to it, functions it can perform and events which influence the way it performs. It is independent of any current or proposed implementation and so represents the logical requirements that successive implementations are designed to fulfil. The data dictionary should be able to support this model.

At this level the data dictionary should be able to record details of:

— entities and relationships of concern to the enterprise;

— processes of interest to the enterprise or carried out by it;

— responsibility for processes, perhaps in terms of the structure of the organisation;

— flows which result from processes or from external entities or events;

— the connections that exist between entities, processes and events.

The data dictionary should be able to record details of different versions recognised as valid at different times or contexts. It is also essential that the dictionary can define the relationships between the logical entities and the corresponding files and records of the implementation.

Implementation Level

The implementation view is the basic source of information about the physical data processing system. It provides data to help establish the design of the system, to prove its correctness and to identify the impact and cost of changes. It is likely to represent a partial implementation of the logical model. It must be logically consistent with this latter model and not exceed it in scope. At this level the data dictionary represents a coherent, centralised library of data about all aspects of the data processing system, enabling all users to have a clear and consistent view.

Two examples of such data are:

Data Description elements. These will describe the different data types and structures used in the system, such as records and files. Elements should be described in terms of their:

Names: including aliases and past names.

Classification: Description, ownership, status, etc.

Representation: Type, length, order, etc.

Use: Frequency and volumes.

Administration: Memory and storage requirements.

Process Description elements. These will also demand the same type of metadata as the Data Description elements. Further information might include:

Program size – in some appropriate metric.

Processing type – Batch or On-line.

Parameters – Number and types required.

Several versions of programs and data structures may exist at any one time. These may represent live, test or design states and this must be recognised and recorded as such.

The data dictionary should also enable descriptions of the implementation-level structures to be established and maintained. This may be achieved through a direct input language, from program data definitions in high level languages, from a DBMS source definition or from program procedure definitions.

The implementation-view also demands details about the physical storage of data and its use. Facilities required include recording of physical attributes such as:

— storage media: storage media type, eg disk

— storage size: describes space requirements, eg 640 Kbytes

— CPU: describes the CPU name and size required.

The data dictionary system should validate input for syntax, consistency and completeness. These checks should include:

— the characteristics of each physical file;

— the contents of each file;

— each physical structure. Checks that all the constituents of the implementation data structure are allocated to at least one physical file.

The implementation view contains all the information necessary to derive an 'optimum' operational schedule. This is supported by the collection of performance and utilisation statistics such as:

— frequency: indicates the average frequency that the file is accessed (such as daily, weekly);

— response: refers to the response time of a process;

— log information: shows statistics on when a record or file is accessed, by whom, and the activity that is performed;

— usage statistics: records summary of usage.

Holding data volumes in each operational definition and the physical description of the files themselves provides information for the realistic simulation of database performance. This gives the facility to tune the performance of the system to achieve a database or file structure that gives optimum performance.

8.2.3   Data Dictionary Functions

The previous section provided a flavour of the contents of a data dictionary as well as introducing performance simulation as one of its possible functions. Similarly, the documentation and control features of a DDS have been described earlier in this chapter. However, the data dictionary has other important functions. These include:

Consistency Checking

This is an essential feature of a data dictionary. In the context of a data flow diagram this can answer such questions as:

— are there any data flows specified without a source or destination?

— are there any data elements specified in any data stores that have no way of getting there, as they are not present in any of the incoming data flows?

— does a process definition demand a data element that does not enter that process?

— are there any data elements in any data flows entering processes that are not used in the process and/or do not appear in the output?

The verification of system consistency is a vital task that eliminates a considerable amount of desk checking. It permits the insertion of vital, but omitted, data elements and the deletion of irrelevant ones. Consequently, the data dictionary not only ensures the validity of the design but also identifies and justifies the role of each data element. Thus it is possible to demonstrate why certain data is collected and where and how it is used.

Testing

The development and entry of test data is extremely time consuming. But if descriptions and ranges of values are already stored then test data can be automatically generated.

Coding

The description of data structures may be detailed enough for the generation of data descriptions in the host language or Data Manipulation Language (DML) through a precompilation pass of the dictionary. Furthermore, if the implementation level model holds information on the sequence in which data is used by processes then automatic program code definition is feasible. A further obvious application lies within the definition of validation rules in the data dictionary permitting the generation of data validation routines and integrity checks.

Change

Program and system maintenance is a major system overhead. One of its most time consuming and difficult tasks is the tracing of the effects of changes through the complete system. Impact Analysis is the term given to the analysis of the effect of proposed program and system changes. The recording of the relationships that exist between the various entities should allow the effects of addition, amendment or deletion of a particular entity to be predicted throughout the whole system. Thus the system and resource implications of a change can be completely understood. It may also be possible to generate some of the coding changes automatically in an active, or potentially active, dictionary.

8.2.4 Reporting and Security

The initial documentation role of many data dictionaries has led to most commercial DDS having flexible and comprehensive reporting facilities.

A variety of analyses will be required and the facility to search for textual descriptions is probably essential if the full potential of the system is to be realised. The output facilities will play an important part in the selection of an appropriate DDS because these are the most immediate part of software. The principles of good output design, described in our companion text *Introducing Systems Design*, have to be observed.

Security codes can be assigned to individual systems and/or metadata entities, or to part or all of the data dictionary. These are used to restrict access to specific meta objects (entities, processes, files, records, etc) and this type of security can be implemented through passwords, authority or ownership facilities. Security design is considered in the companion text and the general principles outlined there apply to the design of data dictionary systems.

8.2.5 Data Dictionary: Summary

Thus the data dictionary is an important tool in managing the corporate data resource. Furthermore, it underlies all the models that have been introduced in the rest of this text, providing the metadata detail required for a successful understanding of current practices and a sound design of their successors. Data dictionary systems have progressed from documentation tools performing basic cross-references to software that provides support for all stages of systems development.

Current data dictionary systems are more active than their predecessors. In a number of cases it is the dictionary that drives data analysis and database design tools, application generators and report writers, with the dictionary facility being used to coordinate these various tools, models and phases. Most Analyst Workbenches (introduced in Chapter 1) rely upon a central dictionary or encyclopaedia.

8.3   ENTITY LIFE HISTORIES

8.3.1   Introduction

The last two chapters have introduced two important modelling tools – data flow diagrams and data analysis. These models are essentially complementary, with the former highlighting the dynamic aspects of data and the latter the static. It would obviously be useful to have a model that ties these two fundamental views together, and the entity life history is a possibility.

The entity life history (ELH) shows what events take place on an entity over time. For example, an ELH of a customer's bank account would show the opening of the account, the transactions that take place (deposits, withdrawals, interest, charges) and the final closing of the account. Particular attention is paid to timing constraints; a customer cannot withdraw money from a bank before opening an account. In general, the ELH examines what causes each entity to be created, deleted and amended during its time in the system. The processes identified in this examination should already be identified on the data flow diagram and so, in this respect, the entity life history acts as a completeness check to the DFD. This double check is particularly useful, because it means that processes have been defined from two perspectives. Firstly, from the dynamic viewpoint of the data flow diagram, and secondly from the static entities of the entity-relationship model.

This chapter introduces the notation, construction and application of the entity life history. A general approach is taken in an attempt to show the basic principles of the model. ELHs and their equivalents are likely to have a much more detailed notation in specific methodologies which are attempting to link two particular data and process driven models.

8.3.2   Entity Life History: Notation

The entity life history can be shown as a tree diagram of boxes. The top box is called the 'root' and the boxes with no connections below them, the 'leaves'. Any parent box can have one or more children, but no child can have more than one parent. An example tree diagram is given in figure 8.2.

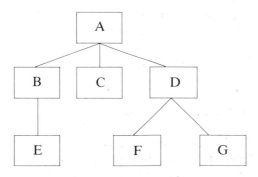

Figure 8.2   Entity Life History: Structure

Three types of activity can be represented on the ELH:

Sequence: The boxes are left unmarked and each event occurs only once in a sequence that is read from left to right. Daily activities are an obvious example. No other sequence is permitted for an occurrence of this entity.

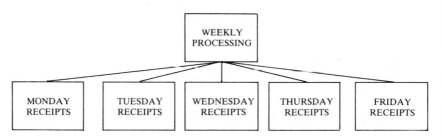

Iteration: An iteration is shown by marking the child box with an asterisk in the upper right hand corner. Each iteration can have only one child. Hence the iteration of month has the child day which is marked with an asterisk.

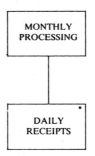

Selection: A selection is indicated by marking each child box with a small circle in the top right hand corner. A selection can have two or more children but for each instance of the parent box only one of the children must occur. Thus a RESULT in an examination must be a PASS, a FAIL, or a REFERRAL. If 'nothing' might happen, then a null box may be included.

Entity Life History: Construction

For each entity:

1   List the events that affect the entity. In a very constrained example.

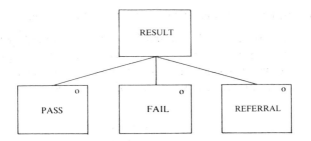

Entity: ACCOUNT
Action: OPEN-ACCOUNT, TERMINATE, PAY-IN, WITH-DRAW.

There are three main types of event:

— Events based upon time or time-cycles.

— Events which are caused by changes occuring outside the system.

— Events which reflect changes within the system.

2   Order the actions in the sequence that they must take place:

OPEN-ACCOUNT, PAY-IN, WITHDRAW, TERMINATE

3   Identify which actions are alternatives (Selection).

PAY-IN and WITHDRAW are alternative transactions. Therefore we are aware that these must be children of the same box. TRANSACTIONS seems a reasonable term for this new parent box.

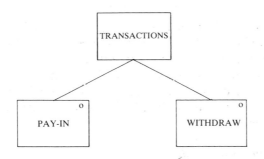

4   Identify which actions take place more than once (Iteration):

The iterative events are PAY-IN and WITHDRAW which we have

represented as children of TRANSACTIONS. Thus TRANS-
ACTIONS is an iterative event and so is the child of a new box which
can be called ACCOUNT-BODY.

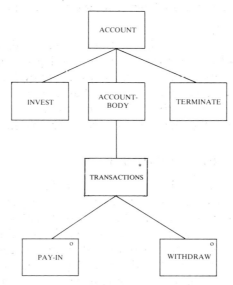

8.3.3   Entity Life History: Uses

1   To check that all events which modify the data have been considered
    in the processes of the data flow diagram and in the transactions that
    support the construction of the entity-relationship model.

2   To analyse what happens if an event occurs out of sequence or an
    unexpected termination takes place. This can be used for specifying
    procedures for error trapping or for handling exceptional occur-
    rences. For example, in the banking model, WITHDRAWS are
    unlikely to be permitted unless an INVEST action has taken place.

3   To identify attributes that have not been uncovered in previous
    models. Neither the data flow model nor data analysis cope
    particularly well with time and sequence. Thus important insights
    may emerge from the use of a model that stresses order and
    chronology.

4   To support some of the decisions that have to be made in the second
    level data design stage in the companion text – *Introducing Systems
    Design*.

5   To clarify relationships between entities and show the paths that may have to be navigated in a subsequent design. This may demand the insertion of new relationships to permit ease of navigation in the implemented data model. This again is covered in more detail in the companion text.

An example of two possible entities is given in Figure 8.3, where the CUSTOMER structure is an interleaving of the ACCOUNT structure. A very similar analysis to this is used by Jackson (Jackson, 1983) to show how this also contributes to the clarification of entities. Thus a mis-specification of entities in the entity model may be unearthed through the construction of a set of entity life histories.

8.3.4   Entity Life Histories: Summary

The notation and examples in this section have been based upon Jackson's Entity Structure Step which is a central part of his systems development approach (Jackson, 1983). However, although this technique has been taken out of its context, we feel that the first two stages of the Jackson approach (entity actions and structure) can be used as a coherent support model for the other two views of the enterprise. This is because it stresses aspects of the system that both data flow and data analysis are poor at modelling.

In practice, problems can be caused by the term entity. Jackson is quite clear that an entity is a real-world occurrence. He gives the following general criteria for entities:

1   An entity must perform or suffer actions, in a significant time-ordering.

2   An entity must exist in the real world outside the system, and must not be merely a part of the system itself or a product of the system.

3   An entity must be capable of being regarded as an individual, and, if there is more than one entity of a type, of being uniquely named.

It is therefore different from the meaning used in database methods, where an entity may have no real world actions, either suffered or performed. (Jackson, 1983.)

In general, Jackson models have fewer entity types than a data analysis model and this must be recognised when constructing an ELH. Not all entities on the data analysis model will have relevant histories

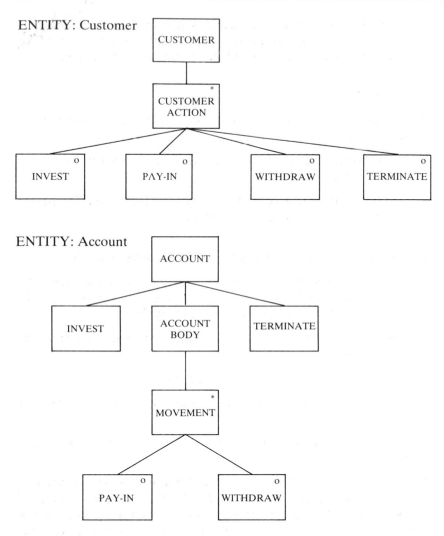

Figure 8.3   Interleaving ELH Structures (from Jackson, 1983)

(indeed the ELH of such entities are usually trivial) because they do not exist in the real world.

However, despite this last point, the entity life history provides a further perspective on the data of an organisation. We now have three important complementary models that describe the logical system from three different perspectives. Together, supported by a data dictionary, they go a long way to providing a comprehensive view of the information system operations and requirements of the organisation.

8.4 SUMMARY

This chapter has provided two brief tutorials in important supporting tools. In doing so it has:

— Introduced the possible role, scope and content of a data dictionary.

— Outlined the purpose of the entity life history and described a general notation for its construction.

8.5 PRACTICAL EXERCISES AND DISCUSSION POINTS

1   One of the problems of examining data dictionary software is the wide range of facilities and nomenclature used by the vendors. By using a suitable reference book (such as Mayne):

— Define what six different packages call the concept of the meta-entity.

— Assess the activity of six different packages.

2   Discuss the problems of introducing a data dictionary system into an organisation.

3   List four advantages of using an active rather than a passive data dictionary.

4   Investigate the first two stages of Jackson Systems Development – the entity action and structure steps. Redraw the entity diagram of the previous chapter as a Jackson entity diagram. Assess the ease of construction of this diagram and explain the differences between the two models.

5   Write the English narrative that describes the rules and activities modelled by the ELH given in Figure 8.4 on page 228.

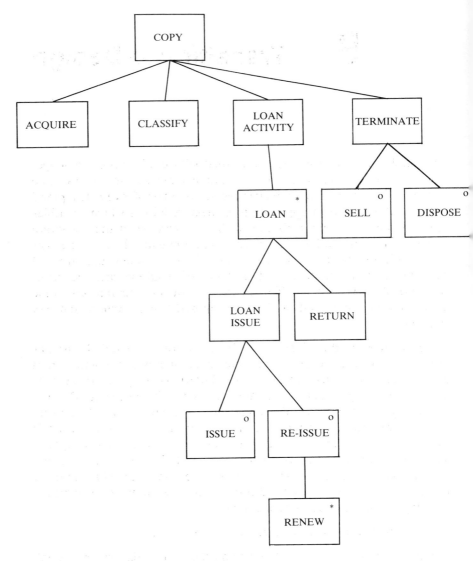

Author: Gillian Mills, Leicester Polytechnic

Figure 8.4   Illustrative Entity Life History

9 Transition to Design

This book has primarily been concerned with establishing the scope, operations and requirements of information systems development. The system requirements have been defined at a number of levels. Chapter 2 introduced a top-down approach to project definition which identifies opportunities for computerisation within the context of the corporate needs of the organisation. The business requirements given priority are then subjected to a Feasibility Study. This examines the area of application in the context of three types of requirements: technical, operational and financial. The Feasibility Report summarises these system requirements and outlines a proposed logical system that will fulfil them.

Acceptance of this Feasibility Report leads to detailed analysis (Chapter 5) where the operations and requirements of the application area are considered in more depth. New knowledge may lead to changes in the system requirements originally specified in the Feasibility Report. Increasing familiarity with the application area will permit all requirements to be stated with more confidence and detail. This precision is increased by the logical models described in the last three chapters. These will provide the basis for the physical design that follows in the companion text. In some respects Chapter 8 began to introduce some of these considerations. The data dictionary provides valuable support for both logical and physical system design.

Design will primarily be concerned with five aspects:

Files or databases: Based upon the entity-relationship model and supported by the entity life history and the data dictionary.

Processes and programs: Based upon the processes of the data flow diagram and supported by the entity life history and data dictionary.

Inputs and Outputs:	Based upon the flows of the data flow diagram and supported by the data dictionary.
Interfaces:	Underpinned by understanding of requirements and skill levels and explored within the context of the data flow diagram.
Controls:	Supported by the data dictionary.

This text has concentrated upon:

— The modelling of current information systems and determining user requirements. This may be supported by Prototyping.

— The modelling of logical systems that underpin both current and planned physical implementations. This may be supported by automated tools.

The context of the system development has been validated by the top-down approach to project selection and feasibility. The implementation of the logical model is the prime concern of the companion text.

Bibliography

CHAPTER 1

Alavi M, An Assessment of the Prototyping Approach to Information Systems Development, *Communications of the ACM,* June 1984, vol 27, no 6

Boehm B, Prototyping Versus Specifying: A Multiproject Experiment, *IEEE Transactions on Software Engineering,* May 1984, vol SE-10, no 3

Er MC, Classic Tools of Systems Analysis – Why They Have Failed, *Data Processing,* December 1986, vol 28, no 10, pp 512–513

Lee B, *Introducing Systems Analysis and Design,* vol 1, NCC Publications, 1978

Lobell R, *Application Program Generators: A State of the Art Survey,* NCC Publications, 1984

Martin J, *Application Development Without Programmers,* Prentice-Hall, 1982

Yourdon E, Whatever Happened To Structured Analysis? *Datamation,* 1 June 1986, pp 133–138

CHAPTER 2

Ackoff R, *A Concept of Corporate Planning,* Wiley Interscience, 1970

Checkland P, *Systems Thinking, Systems Practice,* Wiley, 1981

Grindley K, Humble J, *The Effective Computer,* McGraw-Hill, 1973

Hussey D, *Corporate Planning – Theory and Practice,* Pergamon, 1982

Wilson B, *Systems: Concepts, Methodologies, and Applications,* Wiley, 1984

Wood-Harper A, Antill L, Avison D, *Information Systems Definition: The Multiview Approach,* Blackwell, 1985

Wood-Harper A, *A Distance Learning Project at South Bank Polytechnic*

CHAPTER 3

Chapin N, Economic Evaluation in Systems Analysis and Design – a Foundation for the 1980s, Cotterman W *et al,* eds, North-Holland, 1981

Collins G, Blay G, *Structured Systems Development Techniques: Strategic Planning to System Testing,* Pitman, 1982

Lumby S, *Investment Appraisal,* Nelson, 1981

Parkin A, *Systems Analysis,* Edward Arnold, 1980

CHAPTER 4

Autosate, *Communications of the ACM,* July 1964, vol 7, no 7, pp 425–432

Cherry C, *On Human Communication,* MIT Press, 1978

Daniels A, Yeates D, *Basic Training in Systems Analysis,* Pitman, 1971

Groner C, Hopwood M, Palley N, Sibley W, Requirements Analysis in Clinical Research Information Processing – A Case Study, *Computer,* September 1979, vol 12, no 9, pp 100–108

Hein K, Information System Model and Architecture Generator, *IBM Systems Journal,* 1985, 24, 3/4, pp 213–215

Kimmerly W, Restricted Vision, *Datamation,* 6 March 1984

McMenamin S, Palmer J, *Essential Systems Analysis,* Yourdon Press, 1984

Moser C, Kalton G, *Survey Methods in Social Investigation,* Heinemann, 1971

Parkin A, Thornton S, Holley P, Can Fact-finding Be Automated? *Automating Systems Development,* Benyon D, Skidmore S, Plenum Press, 1987

Rosensteel G, Why Systems Analysis Training Fails, *Computer World,* November 1987

Skidmore S, Sitek J, Are Structured Methods for Systems Analysis Being Used? *Journal of Systems Management,* June 1986, pp 18–23

Yourdon E, Whatever Happened to Structured Analysis? *Datamation,* 1 June 1986

CHAPTER 5

Fergus RM, Decision Tables – What, Why and How, *Proceedings College and University Machine Records Conference,* University of Michigan, 1969, pp 1–20

Gane C, Sarson T, Structured Systems Analysis, *Improved System Technologies,* 1980

Grindley C, The Use of Decision Tables Within Systematics, *Computer Journal,* August 1968, pp 128–133

Lew A, Proof of Correctness of Decision Table Programs, *Computer Journal,* 1984, vol 27, no 3, pp 230–232

Martin J, McClure C, *Diagramming Techniques for Analysts and Programmers,* Prentice-Hall, 1984

NCC, *Student Notes on NCC Data Processing Documentation Standards,* NCC Publications, 1979

NCC, *Standards for Structured Systems Analysis and Design,* NCC Publications, 1987

Parkin A, Thornton S, Holley P, Can Fact-finding Be Automated? *Automating Systems Development,* Benyon D, Skidmore S, Plenum Press, 1987

CHAPTER 6

Beer, Stafford, the quote comes from a METRA paper called *Love and the Computer,* 1964, vol 111, no 1. Parts of Beer's paper may be found in *The Manager,* October and November 1963

de Marco T, *Structured Analysis and System Specification,* Prentice-Hall, 1979

Gane C, Sarson T, *Structured Systems Analysis,* Prentice-Hall, 1979

NCC, SSADM Manual, 1987

Sumner M, Sitek J, Are Structured Methods for Systems Analysis and

Design Being Used? *Journal of Systems Management,* June 1986, pp 18–23

Yourdon E, Whatever Happened to Structured Analysis? *Datamation,* 1 June 1986

CHAPTER 7

Albano *et al, Computer Aided Database Design,* North-Holland, 1985

Avison D, *Information Systems Development: A Database Approach,* Blackwell Scientific Publications, 1985

Chen P, The Entity-relationship Model: Toward A Unified View of Data, *ACM Transactions on Database Systems,* March 1976, pp 9–36

Date C, *An Introduction to Database Systems,* vol 1, 4th edition, Addison-Wesley, 1986

Holley P *et al,* Automated Aids for Entity Relationship Modelling and Information System Design: An Annotated Bibliography, *CARIS Working Paper no 16,* School of Mathematics, Computing and Statistics, Leicester Polytechnic, 1987

Howe D, *Data Analysis for Data Base Design,* Edward Arnold, 1983

Kent W, A Simple Guide to Five Normal Forms in Relational Database Theory, *Communications of the ACM,* February 1983, pp 120–125

Veryard R, *Pragmatic Data Analysis,* Blackwell Scientific Publications, 1984

CHAPTER 8

BCS, *British Computer Society: Data Dictionary Systems Working Party,* BCS, 1977

Jackson M, *Systems Development,* Prentice-Hall, 1983

Leong-Hong B, Plagman B, *Data Dictionary/Directory Systems: Administration, Implementation and Usage,* Wiley, 1982

Mayne A, *Data Dictionary Systems: A Technical Review,* NCC Publications, 1984

Plagman B, Moss C, *Alternative Architecture for Active Data Dictionary/ Directory Systems,* Auerbach, 1978

Redfearn P, The Role of the Corporate Dictionary, *Automating Systems Development*, Benyon D, Skidmore S, Plenum Press, 1987

Skidmore S, Wroe B, *Introducing Systems Design*, NCC Publications, in Press

Index